Realistic and Ethical Use of Artificial Intelligence

Current perspectives on artificial intelligence (AI) tend to focus on the wide parameters between fear and hope. *Realistic and Ethical Use of Artificial Intelligence* provides some (ethical) hope with realistic solutions. It deals with the complexities of AI, unpacking specific challenges presented by the AI problem and suggesting what we do about them.

This book is written by an experienced researcher and a young innovator, with expert reflections from researchers and practitioners. It promotes community-focused, collective values and ethical uses of AI, and it challenges ideologies based on capitalist modes of consumption, privilege and exploitation. Chapters explain current understandings of AI, how it has developed since its inception and the various types of AI. The text explores issues such as scraping of data, privacy, unethical corporate practices, trust and AI abuse, as well as outlining the potential environmental, social and psychological impact of AI – for better and for worse. Discussions across this book are underpinned by both research and media perspectives that introduce concerns and offer suggestions and explanations about ways to use AI in realistic and ethical ways.

This up-to-date text is ideal for anyone looking for a deeper understanding of what counts as AI, the moral questions troubling both researchers and the general public and how AI can be a fantastic tool if built and used ethically.

Maggi Savin-Baden is a Professor and Senior Research Fellow at Las Casas Institute for Social Justice Blackfriars Hall, University of Oxford. She has authored, coauthored and edited 29 books in the areas of innovative learning, digital fluency, artificial intelligence, the postdigital, digital afterlife, pedagogical agents, qualitative research methods, problem-based learning and the metaverse. She is currently co-editor of the Metaverse book series. In her spare time, she runs, bakes, climbs, does triathlons and wild swimming.

Zak Savin-Baden completed his degree in Computer Science from Birmingham City University, gaining a first. He has since completed his master's degree in Artificial Intelligence, published an article and 2 book chapters as well as providing introductory talks on AI, and he is currently working towards a PhD. In his spare time, he is a runner, gamer, programmer and enjoys making 3D models with his 3D printer.

Realistic and Ethical Use of Artificial Intelligence

Maggi Savin-Baden and Zak Savin-Baden

CRC Press
Taylor & Francis Group
Boca Raton London New York

CRC Press is an imprint of the
Taylor & Francis Group, an **informa** business
A CHAPMAN & HALL BOOK

Designed cover image: Shutterstock

First edition published 2026
by CRC Press
2385 NW Executive Center Drive, Suite 320, Boca Raton FL 33431

and by CRC Press
4 Park Square, Milton Park, Abingdon, Oxon, OX14 4RN

CRC Press is an imprint of Taylor & Francis Group, LLC

© 2026 Maggi Savin-Baden and Zak Savin-Baden

ISBN: 978-1-041-03241-0 (hbk)
ISBN: 978-1-041-03236-6 (pbk)
ISBN: 978-1-003-62294-9 (ebk)

DOI: 10.1201/9781003622949

Typeset in Palatino
by codeMantra

Ignorance is the curse of God; knowledge is the wing wherewith we fly to heaven.

Shakespeare (1600)

For Randi Slack, friend and editor extraordinaire

and

Megan Spinks and Rooda Sheik Ahdmed, friends and fellow students

Contents

Figures

Tables

Acknowledgements

This book has been both a joy and a challenge to create in a rapidly changing landscape of artificial intelligence in a postdigital world.

We are grateful to the following experts who have provided insights into their work that serve to illuminate some of the ideas and issues we have discussed in the chapters: Chapter 1 Professor Beth Singler, University of Zurich, Switzerland; Chapter 2 Mike Savin National Energy System Operator, UK; Chapter 3 Dave White, University of the Arts, UK; Chapter 4 David Burden, Daden Limited, UK; Chapter 5 Wesley Goatley, University of the Arts, UK; Chapter 6 Professor Nigel Crook, Oxford Brookes University, UK; Chapter 7 Claddagh NicLochlainn, Institute for the Future of Work, UK; Chapter 8 Kester Brewin, Institute for the Future of Work, UK; Chapter 9 Sarah Hayes, Bath Spa University, UK; and Chapter 10 Dr Elaine Kasket, UK

We are also thankful to John Savin-Baden who has supported the editing process and managed and merged our voices across the texts. Any mistakes and errors are ours.

Biographies of experts

Kester Brewin is Associate Director at the Institute for the Future of Work, a research and development institute exploring how AI and automation technologies are impacting the labour market in the UK. He is also an author of several books on technology and social history, including, most recently, *God-like: a 500-Year History of Artificial Intelligence in Myths, Machines, Monsters*

David Burden has been creating conversational AIs and virtual realities since the 1990s. David has published widely and is currently series co-editor for Taylor & Francis on The Metaverse Series. David is an ex-British Army officer, a Chartered European Engineer and is currently studying for a PhD in Wargaming Urban Conflict.

Nigel Crook is Professor of AI and Robotics, Associate Dean for Research and Knowledge Exchange and Director of the Centre for AI, Culture and Society at Oxford Brookes University. He graduated from Lancaster University with a BSc (hons) in Computing and Philosophy in 1982. He has a PhD in explainable intelligent machines and 40 years of experience as a researcher in AI. His research interests include biologically inspired machine learning, social robotics and moral machines

Wesley Goatley is a critical artist and researcher based in London, UK. His work examines artificial intelligence technologies and their relations to society, geopolitics, and the climate crisis, and how art practice can intervene and explore these tools and topics. He has given talks on his practice and research at international events such as Milan Design Week, Global Art Forum Singapore, CTM Festival Berlin, and the European Data Forum in Eindhoven. His installations, performances and films have been shown at international venues such as Eyebeam in New York, Berghain in Berlin, The Nam June Paik Art Center in Seoul and the Victoria & Albert Museum in London.

Sarah Hayes is Professor of Education & Research Lead in the School of Education at Bath Spa University. Sarah's research includes linguistic analysis of Higher Education policies and examining society through a postdigital lens. Her books include *The Labour of Words in Higher Education* (2019), *Postdigital Positionality* (2021) and the EPSRC funded *Human Data Interaction, Disadvantage and Skills in the Community*. Sarah has taught Sociology, Education and Computing and is an Associate Editor for the *Postdigital Science & Education* journal.

Dr Elaine Kasket, Visiting Professor at Centre for Death and Society at University of Bath, is a psychologist, cyberpsychologist, and author of *All the Ghosts in the Machine: The Digital Afterlife of Your Personal Data*. She writes and speaks widely on the experience and ethics of digital legacy, AI and posthumous identity.

Claddagh NicLochlainn is a researcher and policy professional specialising in AI governance and tech regulation. She leads parliamentary affairs at the Institute for the Future of Work and is a Newspeak House Fellow. Claddagh holds a Master of Public Policy from the Blavatnik School of Government, where she focused on legal and ethical frameworks governing the use of AI in border control. She writes for a number of different outlets, exploring the intersection of technology, law and structural inequality.

Mike Savin is a senior IT leader with National Energy System Operator enabling business transformation and turnaround of failed programmes. Since gaining an MBA from the Open University, he has become a successful Head of Delivery for over 25 years and his industry experience includes oil and gas, (first LNG to Queensland), telecoms, renewables energy sector and developing cyber security capabilities. In his spare time, he enjoys climbing, skiing, cycling and running.

Beth Singler is the Assistant Professor in Digital Religion(s) at the University of Zurich, Switzerland, where she is also the Co-Director of the University's Research Priority Programme (URPP) in Digital Religion(s), a member of the Directorate of the Digital Society Initiative and co-lead of the Media Existential Encounters and Evolving Technology Lab. An Anthropologist, her internationally renowned research focusses on the entanglements of religion and AI as well as the social and ethical implications of the current age of AI.

David White works at the intersection of education, research and digital innovation. He is known for developing the Digital Visitors and Residents continuum. In his role as Dean of Academic Strategy at the University of the Arts, London, he oversees the academic aspects of fully online provision. He is a Principal Fellow of the Higher Education Academy and President of the Association for Learning Technology.

Introduction

Maggi In the autumn of 2024, we were standing at the doorway of Blackfriars Hall, Oxford reflecting, as the sun set over Jericho, on a talk I had just given on academic integrity and artificial intelligence. The evening seminars run throughout term time, and after the seminar, there is wine and pizza and then time for questions. The debate and discussion after the talk was vibrant, and as we stood outside in the twilight, we realised that perhaps a book was required that dealt with the complexities of AI. Our thinking was confirmed when we were asked to run a workshop for ministers and church leaders in the Midlands, many of whom did not realise how much of their data was being tracked and used, never mind the unethical and illegal scraping of data being carried out by Large Language Models. We pitched the idea to Randi Slack with whom I had done a number of books, and she felt the combination of younger and older perspectives on this issue would be a sound combination. I tend to write in an academic style, Zak in a more accessible but no less rigorous way. As we have travelled through the journey of this text, we have discussed, debated and disagreed (not often) but what we believe we have created is an up-to-date text, under pinned by both research and interesting media perspectives, that will introduce concerns and offer suggestions and explanations of ways of using AI in realistic and ethical ways.

Zak The agony of being an AI scientist in this day and age is that you look around and find that the big tech corporations are pushing ahead with AI, often with unethical practices. Throughout both my degrees, I was taught how to build AIs using data that was either open source or in the public domain and free to use. However, when you turn to the industry, the Big Tech corporations seem to be stealing every bit of data that isn't nailed down. The General Public, however, are taking to these models like a fish to water, asking ChatGPT for advice, making all sorts of fun scenarios and generating all sorts of crazy images. Then, there is the advertising. Whenever I open YouTube, I'm bombarded with Grammarly ads and companies yelling about how AI revolutionised their business. If you need AI to make a business good, is it a good business to start with? What do we do about this AI problem? The companies pressing ahead with this technology are more worried about whether they can rather than whether they should. Therefore, the idea of this book is to unpack some of the specific challenges presented by the AI problem and to suggest what we do about it. It's been a long journey but a fun one, and I think the main takeaway from everything going on is that AI is a fantastic tool, provided it is built ethically.

DOI: 10.1201/9781003622949-1

Overview

This book begins in the first chapter by exploring current understandings of AI and then discusses how it has developed since its inception at a workshop held on the campus of Dartmouth College in 1956. It will provide an overview of the various types of AI, such as early understandings of AI, and then present simple algorithms, Generative AI and Neural Networks, using accessible tables. The second section of the chapter explores issues of privacy in order to illustrate the privacy threats we face from many different arenas. The final section of the chapter examines different forms of capitalism and then analyses how these are affected by AI and their impact on society in general. Then, in Chapter 2 we discuss AI in a more in-depth way than Chapter 1, with the first section explaining the specifics of algorithms and Big Data, and their various implementations and challenges. The second section explains the relationship between AI, Machine Learning, Neural Networks and Deep Learning, and in particular how Neural Networks become Deep Learning. The final section explores some of the problems that emerge during AI development.

Chapter 3 examines Large Language Models and Generative AI and analyses some of the facts and myths related to Large Language Models in terms of whether they do 'understand' or only function on the basis of statistical occurrence models. It then explores understandings of trust in relation to Large Language Models, particularly ChatGPT, as well as the extent to which people do in fact trust them. Large Language Models are currently treated with too much respect and hold unjustifiably high reputations for being what are essentially glorified guessing machines; thus, the final section examines issues of theft. From this, Chapter 4 analyses the uses and dangers of LLMs, examining concerns, promises and issues of social justice, and then examines the impact of these models on humanity and society. It offers an overview of some of the recent cases of AI abuse and suggests ways in which it might be possible to anticipate the future and find ways of developing methods to minimise the effects of such abuse. Overall the chapter argues that while LLMs are a fascinating tool, users must remember that they are only a tool and should be treated as such.

Chapter 5 moves into the broader sphere of AI and climate change, exploring the environmental, social and psychological impact of AI. It then examines power consumption, water consumption, CO_2 emissions to AI and then covert surveillance. The final section will examine the drive towards nuclear options and then analyse the positive and valuable uses of AI which may also minimise environmental effects. It explores the market impact of AI and makes recommendations about what corporation and government responses should be and suggests models and mechanisms that should be put in place. From here, Chapter 6 discusses the morals in the machine, suggesting that the use of AI will always be fraught with moral challenges and great care must be taken to ensure moral data collection, moral AI development and deployment. This chapter explores the idea of morality when regarding AI models and their development and provides an overview of how immorality can creep into AI development and use. The second section of the chapter explores Just War theory, suggesting that the use of AI drones and robotic soldiers needs careful consideration. The idea of morality within the world of AI has always been troubling to researchers, developers and data scientists throughout the technology's development, and so the final section argues that government stances towards AI are underdeveloped, and more work is required in this area.

Chapter 7 examines ethics. The difficulty with the area of ethics and AI is that the AI landscape is continually on the move. A further concern is that dupery and lying have become increasingly part of the political canon so that the values of the good society – the values of equality, democracy and sustainability – are continually being eroded. This chapter begins by exploring what it means to be human and then reviewing the landscape of ethics in the 21st century. It then suggests practical responses to ethical concerns and explores ways of dealing with ethics and AI. The final section of the chapter explores the issues of ethics and AI in relation to the law and the impact of new legislation in 2025. Having analysed and discussed the complexity of AI ethics, Chapter 8 then examines what some of the realistic solutions might be. The overarching argument of this chapter is that the perspective towards AI needs to change; from a corporate perspective, humans should be prioritised in terms of employment, and AI must be implemented in ways that are ethical. In terms of the users, people need to be more cautious about their level of use of AI,\ as well as ensure that their use of AI is ethical. The chapter begins by presenting the serious mistakes some companies have made in publicising their increasing commitment to AI. The second section discusses the user of the AI systems themselves, such as the ignorance and negligent nature of users as well as what the model user should do in response to the world becoming more AI-driven. The final section discusses AI poisoning techniques for use as defence against AI systems.

The final two chapters have more of a future focus, with Chapter 9 exploring the way in which future predictions are resulting in new forms of panic. It begins by examining just what kind of panics we are dealing with and then compares this with predictions for AI and human expertise. The second section of the chapter explores forms of dissent; these forms of dissent illustrate the impact of AI on academic discourse and political, social and digital dissent. The final section explores dissenting acts and the ways in which AI helps and hinders dissent in different spaces. Chapter 10 then explores death, ethics and artificial intelligence. Death remains a taboo subject across the globe, and yet since the early 2000s, there has been a rise in advances and use of death tech. More recently, particularly following the COVID-19 pandemic, death and the digital have become increasingly intertwined. This chapter begins by exploring the death practices and the impact of AI, including the linking of AI to the brain. The chapter then moves on to examine the relationship between transhumanism and AI and then presents some of the current digital afterlife options. The final sections of the chapter explore death ethics and AI and some of the ethical conundrums related to this, including death prediction and AI, death and media waste and digital engagement with the dead trends.

In conclusion

Thus, for us, the argument of this book is that the challenges of AI largely relate to the fact that many definitions are broad and unclear. Current perspectives tend to focus on the wide parameters between fear and hope. Fear in terms of loss of work, enslaving humans and making them obsolete and the increase in surveillance. Hope in relation to providing virtual human support to elderly and disabled people and providing entertainment as well as supporting security and defence systems. In the end, we hope we leave you with some (ethical) hope.

Maggi and Zak, June 2025

1

The AI problem

Introduction

This first chapter begins by exploring current understandings of AI, then discusses how it has developed since its inception at a workshop held on the campus of Dartmouth College in 1956. It will provide an overview of the various types of AI, such as early understandings of AI, and then present simple algorithms, Generative AI and Neural Networks, using accessible tables. The second section of this chapter explores issues of privacy in order to illustrate the privacy threats we face from many different arenas. The final section of this chapter examines different forms of capitalism and then analyses how these are affected by AI and their impact on society in general. The overarching argument of this chapter is that since AI in all its forms is developing rapidly, there needs to be clarity in schools, businesses, companies and higher education about the different types of AI and how these might be managed ethically and effectively.

Delineating AI

With the rapidity of change in this field, delineating the different types is a complex business. In this chapter we offer a general overview and then provide a more in-depth analysis in later chapters. It is notable that artificial intelligence (AI) in the past has been grouped broadly into three main types. First, Deep AI, which involves creating virtual humans and digital immortals with sentience: the ability to feel, perceive, or experience. Second, pragmatic AI, where machines are used for tasks such as mining large data sets or automating complex or risky tasks. Third, Marketing AI, where companies use algorithms to anticipate the customer's next move and improve the customer journey. Often when AI is spoken about in the 21st century, the perspectives that many people have gained are from information presented by media and film. Some of the early versions were Replicants in Blade Runner (Deeley & Scott, 1982) which provides an image of an Android who becomes independent and unpredictable

DOI: 10.1201/9781003622949-2

and the problems of knowing how to spot them. In 2001, A Space Odyssey Hal is an AI in control of a spaceship who prioritises the mission over the crew so that we see that goals that were set for an AI have unintended consequences. These early versions tended to be quite chilling, but in Star Trek (Berman, 1995) a sentient Android was created as a crewmate, but it was not always clear whether it had emotions or was too artificial. However, Ultron in Avengers: Age of Ultron (Feige & Whedon, 2015) is sentient rather than being created as a digital immortal persona of other people and was portrayed as sympathetic, supportive and often eminently reasonable. There have also been several sinister portrayals such as Will in Transcendence (Kosove et al, 2014) where his consciousness is placed into a quantum computer with the idea of the possibility of a superhuman AI being who can transcend the laws of physics – and who also behaves in an erratic and dangerous way. Recchia (2020) analysed a dataset of over 100,000 film subtitles and identified control (or loss of it) as a recurring motif in films about artificial intelligence, which tends to reflect broad public opinion about AI in general (Burden & Savin-Baden, 2019).

AI, Machine Learning and other derivatives of AI tend to be conflated, largely by the media, because they are defined in ways that would seem to imply that they are the same thing. For example Machine Learning is a computer system that can learn to make decisions based on the examination of past inputs and results. Whereas simple algorithms are just the use of algorithms to predict behaviour such as shopping patterns. There are other forms of AI that tend to be much more technical, such as Deep Learning and other more unorthodox, and as yet, unviable options based on assumptions about the possibilities for uploading the brain. Whilst some of these might be possible in the future, the notion of reanimating a brain from a cryopreserved body based on mind files (pictures, videos and documents in a digital archive) created predeath seems unlikely. However, the rapid expansion of AI, with almost daily developments occurring, means that the speed of change can result in little or no legal compliance. There is little understanding of the differences in the types and the impact of AIs and the creation of faulty or biased applications. Whilst some organisations are seeking to develop ethical AI frameworks it is clear that these are relatively easy to circumvent, and progress is so rapid that this ideal is largely unmanageable. However, with the rapid advances there has been a shift away from defining AI as 3 main types to much more in-depth and detailed analyses.

Forms of AI

As aforementioned there has been a shift away from using AI to refer to all types of AI and instead the desire to be more specific about terminology. AI is, in the main, based on foundation models that have been trained on a vast corpus of data (invariably stolen) from books, websites and social media. Large Language Models (LLMs) which support ChatGPT and DeepSeek, are the first iteration of these but Driess et al. (2023) note input and output modalities are also being developed.

One specific change in recent years is in the growth, proliferation and discussion of LLMs. LLMs are generative mathematical models that contain a large corpus of text; examples of these are ChatGPT and DeepSeek. Such models are termed 'generative' because we can ask some questions of them, be they very specific ones.

Shanahan (2024) points out that it has become common to use the term LLMs for the generative model itself on the systems in which they are located. The generative models, he explains, develop statistically likely sequences of words so that the system is essentially a dialogue management system. LLMs can also be developed in conjunction with vision language models; examples of these are Flamingo and VilBERT which use multimodal architecture that comprehends image and text data modalities at the same time.

The way in which generative modes work is that when we ask them a question, we are actually asking:

> Given the statistical distribution of words in the vast public corpus of (English) text, what words are most likely to follow the sequence, 'The first person to walk on the Moon was...' A good reply to this question is 'Neil Armstrong'.
>
> *(Shanahan, 2024, p. 70)*

The reason this is important is that recognising that this is a mathematical model prevents us from anthropomorphising a large language model; it is the idea that people speak as if their phone knows where they are, or their car clock does not realise the time zone has changed. Shanahan (p. 79) notes:

> ...today's LLMs, and the applications that use them, are so powerful, so convincingly intelligent, that such license can no longer safely be applied... The careless use of philosophically loaded words such as "believes" and "thinks" is especially problematic, because such terms obfuscate mechanism and actively encourage anthropomorphism.

LLMs thus do not know anything; they just do sequence prediction; they may contain knowledge, but they have no access to any kind of external reality. However, despite the focus on LLMs, it is important to understand how different terms related to AI are being used and what they mean in practice. The next section offers an overview of some of the many varieties.

Simple algorithms

These are a series of instructions or procedures that collect and encode data and, based on specific calculations, transform data into output that brings about a desired outcome. In daily life, algorithms affect media recommendations people receive, such as reviews of books and equipment and Twitter feeds, as well as more personal areas such as Google and Fitbit tracking. Simple algorithms are used for almost all computer programmes, even Complex Enterprise Resource Planning and Customer Relationship Management Systems, since they are highly linear and predictable.

Symbolic AI

The early days of research into AI, from 1946 onwards, were focused on the concept of symbolic AI. This was the process of creating computer programmes that could 'understand' and manipulate symbols in order to reason – often referred to as the so-called cognitivist approach. Anderson (2003, p. 93) describes cognitivism as 'the hypothesis that the central functions of mind – of thinking – can be accounted for in terms of the manipulation of symbols according to explicit rules'. The pinnacle of the symbolic approach, later referred to as 'Good Old Fashioned AI' – GOFAI, was systems such as Cyc (Lenat & Guha, 1989), which aimed to develop a knowledge base that included a significant percentage of the commonsense knowledge of a human being and began in 1984.

Machine Learning

Machine Learning is a form of AI that enables computers to learn and improve from data without being explicitly programmed by using algorithms to analyse data. Thus, it is seen as a system that can learn to make decisions based on the examination of past inputs and results so that its future decisions optimise some parameter – such as facial recognition. However, complex algorithms are often used as a shorthand term for AI, when in fact AI is much broader and more complex than this.

Neural Networks

Neural Networks are advanced Machine Learning models that simulate the human brain's functions, enabling them to learn patterns from data; thus, it is the idea of teaching computers to process data in a way that is inspired by the human brain. Neural Networks were first proposed in 1944 by Warren McCullough and Walter Pitts, two University of Chicago researchers who moved to MIT in 1952. The function of Neural Networks is to classify and categorise data based on similarities. Neural Networks are often used interchangeably with Deep Learning, but Deep Learning is more complex.

Deep Learning

Deep Learning utilises many hidden layers – hence referred to as deep – and requires large volumes of labelled data and processing power. For example, developing a driverless car needs millions of photos and hundreds of hours of video. Deep Learning is distinguished from Neural Networks on the basis of their depth or number of hidden layers.

Generative AI

This refers to deep-learning models that can generate high-quality text, images and other content based on the data they were trained on. Thus, it uses algorithms to create new content, including audio, code, images, text, simulations and videos. LLMs such as ChatGPT are an example of Generative AI since they can answer questions, recognise a picture or create a short story based on the style of a particular author.

Semantic AI

This is the process of getting AI to understand what things are so it can create its own analysis and deductions. For example, Semantic AI enables machines to interpret the intent of content rather than focusing on the exact words being used so that it uses semantic understanding of words such as 'rock' and 'rock music' to guide a user's search. On platforms such as Amazon, it is used to improve product search, but it also prompts recommendations to encourage buying.

Robotics

Whilst robotics is the creation of physical robots, AI is used to programme them to carry out a series of actions autonomously or semi-autonomously. For example, Sycan is an LLM embedded in a robot that is pretrained to perform actions from a data set (Ahn et al., 2022). A somewhat more staged version was when, in the autumn of 2017, Sophia, a humanoid robot, gave a speech at the United Nations. Other examples include vacuum cleaners, toys, drones, NASA's Robonaut, the humanoid and devices such as the robotic exoskeletons that can be used to enable a paralysed person to walk again.

Table 1.1 provides an overview of the challenges of AI with examples.

Artificial intelligence is progressively affecting every area of life, making it feel increasingly uncanny. Mori's concept of the 'uncanny valley' (Mori, 1970) relates to how objects which are almost, but not quite human, create more of a sense of unease than those which are decidedly not human. Mori identified that, when considering artificial human forms, humans tolerated industrial robots but did not particularly respond to them emotionally. Humans showed more affinity and empathy for toy robots and good-quality (anime- and manga-style) human-like but still visually artificial toys. However, research also suggests that people find robots wearing clothes and with fake skin to be creepy, but what is clear is that people will interact with avatars and chatbots 'as if' they were human (Burden & Savin-Baden, 2019). However, as AI advances, there are increasing possibilities to integrate different types

TABLE 1.1

Descriptions, challenges and examples of AI

Type of AI	Description	Challenge	Example
Simple algorithms	Simple algorithms are used to undertake a task which in general is straightforward and predictable.	These algorithms often appear sophisticated but are not, so cannot really be classed as AI.	Facebook algorithms designed to advertise to individuals, based on browsing and shopping history.
Symbolic AI	Creating computer programmes that can 'understand' and manipulate symbols in order to reason.	The assumption that programmes can be created to reason. Seen as a bit passé now and often includes simple algorithms.	Autonomous cars use this to make decisions, such as recognising stop signs and traffic lights.
Machine Learning	The use of data and algorithms to imitate the way that humans learn, gradually improving accuracy.	Often used as a shorthand term for AI, when in fact AI is much broader and more complex than this.	Analysis of Big Data sets.
Neural Networks	AI that teaches computers to process data in a way that is inspired by the human brain.	Can be almost impossible to explain how they reach their solutions.	Facial recognition software.
Deep Learning	AI that uses deep and hidden Neural Networks.	Requires large data sets and can be expensive.	Self-driving cars.
Generative AI	Algorithms that can be used to create new content, including audio, code, images, text, simulations and videos.	Trying to get the machine to match human creativity – and issues about machine plagiarism.	ChatGPT, MidJourney, Claude.ai
Semantic AI	Getting AI to understand what things are so it can create its own analysis and deductions.	Creating and maintaining the knowledge/ semantic graphs (mind maps) that typically support this approach.	WikiData, WordNet.
Robotics	The creation of machines, often human-like, that are programmed to act on behalf of humans.	Has to be pretrained to perform actions from a data set and can easily make mistakes.	Sycan Robotic exoskeletons.

of AI and diverse tasks. Some examples of these include advanced virtual assistants, reading medical images and assessing students' work, yet it is also having an impact on privacy and surveillance. Beth Singler offers her perspective on AI.

EXPERT REFLECTION

Beth Singler, *Assistant Professor in Digital Religion(s)*
at the University of Zurich, Switzerland

WHAT AI DREAMS MAY COME?

When teaching about religion and digitality, and religion and AI in particular, there are times when I am concerned that I might be trying to 'teach grandma to suck eggs'. That the students receiving my lectures are already deeply enmeshed in the day-to-day technical realities of AI and what impact it is having on their lives. However, I was reminded of the disjoint between our ideas of AI and the realities of its context and history recently when one of my students, a retired gentleman who might himself have been around long enough to see several forms of 'AI' come and go, was deeply surprised that the term itself actually dates back seventy years or so to a specific Summer research programme held in the 1950s in the US. He thought AI was only really a few years old. Because GenAI applications like ChatGPT have flooded the conversation and our imaginaries when it comes to AI. Such 'chatty' machines have a linguistic fluency that is designed to convince us that this is the smartest thing that we have ever encountered. But further, that it is the smartest thing we have ever dreamed of making. Not so.

Other technological approaches to 'AI' over those seven decades remind us that our paradigms can shift. Our dreams of AI minds are even patterned on older aspirations and hopes for contact with other beings mostly like us – spirits, fairies, aliens and much more.

Through the anthropological lens that my research on AI uses, it's clear to see that we are also struggling to conceptualise the minds of AI that we are imagining. We are granting them certain attributes (even religious ones) when we try to delve into details (such as we can find) of what this rapidly changing technology is and how it works.

I've used the word 'delve' intentionally in the previous sentence, as specific words such as this have been seen as 'Reverse Turing Tests' – clues that prove that the writer is not a human but is or has used a mimicking machine-like ChatGPT. But even before we can use such shibboleths to prove the origin of such outputs, we are already tangled up in a larger discussion about what AI is or isn't. And, just as my writing could now be accused of being AI-generated (it is not, I swear), we are starting to suspect humans around us are not as human as we thought, because our ways of categorising others are likewise being reshaped by the paranoias and concerns we have about AI. Just as we are seeing the machine as more human-like, we increasingly see the human as more machine-like. And perhaps it is those two connected shifts in categories that are a part of what it is still necessary to think about and be educated about when it comes to AI, no matter the level of technical knowledge we think we already have.

Privacy and other concerns

The growing theft of identity and property, as well as numerous scams, has resulted from the loss of privacy, poor social media self-protection and naiveté about the importance of digital privacy. To date, many people do not ensure that they have up-to-date privacy settings on their phones, thus allowing their data and photos to be shared and reused. Few people check and read website terms and conditions, and even fewer reject cookies from websites. The practice of not rejecting cookies means that your browsing activity is tracked, personal information can be stolen and you are more likely to be susceptible to viruses. Further concerns are as follows:

Plagiarism and cheating

The digital has made it easier not only to copy and paste material from websites but also to buy and sell essays. Artificial intelligence has enabled students to take this a step further by using Generative AI to write assignments and even help them to answer questions at university online interviews. As Sharples remarks, there is a certain irony here

> ... it is ironic that antiplagiarism software uses artificial intelligence to assess the originality of assignments and that different AI (like ChatGPT) can be used to get around plagiarism detection software within seconds. The irony is complete when we realise that GPT-3 can write a review of the student's AI-generated assignment on behalf of the teacher via a simple command: "Here is a short assessment of this student essay:"

> *(Sharples, 2022)*

However, plagiarism prevention in universities is now largely ineffective, and it seems that relatively few academics question this. An example of this is the unquestioning use of Turnitin software, which may highlight some inaccuracies, but it also makes mistakes. The algorithms employed by Turnitin use a process of matching text fragments and comparing students' work with this, offering an originality report at the end of it. Thus, as Rudolph argues, it is important to be aware that 'A first AI circumvents a second AI and is assessed by a third AI. All that the humans do is press a couple of keys, and nobody learns anything' (Rudolph et al., 2023). It is important too that the use of such software is questioned by academics so that students and their assessments are not codified and defined by the very particular writing practice supported by Turnitin algorithms.

Legal violation

Despite data protection legislation like the General Data Protection Regulation (GDPR) (2016) in Europe and The Health Insurance Portability and Accountability Act (HIPAA) (1996) in US healthcare, along with other compliance and risk mitigation measures, legal violation still occurs on a large scale. One example was in July 2021 when European data regulators fined Amazon €746 million, arguing that the company's processing of personal data did not comply with the EU General Data Protection Regulation. Regulators and stakeholders increasingly want companies to be open about data usage and sharing. Yet despite this, data control remains a cleverly managed and hidden activity that is difficult to legislate for and prosecute.

Security

Enforcing security now carries a huge financial cost, and therefore, the protection of some areas is favoured over others. The question is what should be left unsecure and what might this mean? Furthermore, it is also important to consider whether there is now really any possibility of privacy or secure identities. It seems our identities can be both stolen and borrowed, as well as even used against us, perhaps not imminently but certainly in our future lives and work. To date, few people ensure their privacy settings on their phone are set correctly, and few people read the terms of condition on websites or reject cookies – to reduce personal tracking. Furthermore, the frequency of scams and phishing – the process of tricking people into giving away personal information or downloading malware – increases daily.

Privacy threats

Whether there can be any kind of real symmetry between people and machines centres on ethical concerns, which is clearly evident in debates about face recognition and face verification. Big Brother Watch, the independent research and campaigning group, exposes and challenges threats to privacy, freedoms and civil liberties amid technological change in the UK.

One of its main campaigns is FaceOff, which challenges the use of automated facial recognition in policing (Big Brother Watch, 2019).

Equity and fairness

The use of artificial intelligence and Machine Learning has become a popular tool for selecting candidates for interview. However, there have been some well-documented cases of discrimination. For example, in October 2018 Amazon had reportedly scrapped a Machine Learning tool for selecting the top resumes among its job candidates because the system discriminated against women. The bias was due to the fact that the tool was trained on a dataset of resumes from previous applicants, who were predominantly male. Other difficulties have occurred in the past when virtual worlds such as Second Life only allowed limited body shapes, sizes and skin tones (Savin-Baden, 2010).

Data breaches

The range of connections, networks and devices is resulting in an increasing risk of security attacks so that invasion of privacy is a concern, particularly in terms of identity theft. Government surveillance and the monitoring of our shopping habits have been taking place for many years, yet the sophistication of these threats has changed. For example, software such as Zeus is a Trojan horse used for malicious and criminal tasks, such as stealing banking and credit card information, as well as malware that tricks users into unwittingly installing it. Other data breaches include the creeping invasion of personal privacy through technological innovation and arguments for safety. For example, staff being 'required' to use their personal mobile phones to ensure three-factor authentication. Staff in the main do not complain or question this and those who do are deemed to be difficult.

Money laundering

Money laundering is the illegal process of making large amounts of money through criminal activity, such as drug trafficking or terrorist funding, which may appear to have come from a legitimate source. It is a process that involves the *placement* of dirty money into the legitimate financial system, then *layering* which conceals the source through multiple transactions and finally *integration*, where the laundered money is withdrawn from a legitimate source. There has been an increasing possibility of money laundering due to the growth of cryptocurrency, which allows layering transactions to hide the origin of money. Money laundering techniques involving cryptocurrencies, Chen explains, involve the use of 'mixers' and 'tumblers' that break the connection between an address (or crypto 'wallet') sending cryptocurrency and the address receiving it (Chen et al., 2022).

Cybersecurity

Generally speaking, cybersecurity includes network information and computer security. In practice, cybersecurity seeks to ensure digital property remains in the correct hands, although this is increasingly not the case. Cybersecurity has become increasingly complicated as a result of the growth of the interconnections between systems and networks. According to Chithaluru et al (2023), the evolution of cyberattacks is continuing to find new and invasive ways to target even the most sophisticated systems. However, a new type of AI is being developed that is human-in-the-loop, which combines human wisdom and machine intelligence. (Zhang et al, 2022). Since AI can process data quickly but may not judge situations accurately, it is important to include humans who can provide judgement in the face of new changes in the cybernetwork.

We argue across this text that there are no longer spaces that are untouched by the digital since everything is imbued with the digital and therefore necessarily AI. Whilst all places could be argued to be digital, not all such places create a sense of presence or of being embodied or of being in the digital space, which includes the impact of AI on various capitalisms.

Capitalisms

The essential feature and focus of capitalism is the requirement to make a profit. Whilst many countries operate a mixed market economy, the growth of capitalism is resulting in a reduction in equal opportunities, and the supply and demand nature of the free market results in increased poverty. However, in the digital age techno-capitalism, cognitive capitalism, bioinformational capitalism and surveillance capitalism have become huge areas of growth.

Techno-capitalism

This is the commodification of knowledge in fast and diverse ways (Suarez-Villa, 2001), resulting in the corporatisation of inventions and innovation. The impact of this is that the creation and reproduction of technology will affect a society's ability to prosper. Creativity and knowledge are vital resources in the development of techno-capitalism. This can be seen in software, microchips and digital media, which have supported the emergence of techno-capitalism and have made it possible for the information society to develop and spread. Central to the idea of techno-capitalism is the idea that science and technology are part of society, and they are subject to the priorities of capitalism. An example of techno-capitalism is when technology companies such as X (formerly Twitter), Instagram and Facebook extract personal data from users and turn this data into informational commodities to make a profit.

Cognitive capitalism

What is central to an understanding of cognitive capitalism is not about capitalism that is cognitive *per se*, but rather about the accumulation of immaterial assets and the branding and patenting of such assets. In practice, this is seen in the extraction of monopoly rents through intellectual property rights such as Microsoft's putting a patent on the scheduling of meetings, resulting in a licence fee on Android phones. Furthermore, as the importance of knowledge has increased and higher levels of education have developed, this has resulted in a subsequent increase in the value of intellectual labour. The result is that the production of wealth is no longer based solely on goods and material assets but also on knowledge and relational activities. The impact of cognitive capitalism is an increase in working hours and a reduction in division between the home and work. However, as Dafermos & Söderberg suggest, there is resistance through pirates and 'hacktivists' who develop new peer-to-peer technologies and file-sharing networks, with the aim of encouraging 'mass defection from the intellectual property regime' (Dafermos & Söderberg, 2009, pp. 54–55), along with opposition from open access movements and groups who seek to oppose the capitalist enclosure of art, technology and science (Bauwens, 2005).

Bio-informational capitalism

Peters describes the political relationship between biology and information as bio-informational capitalism:

> the emergent form of fourth or fifth generational capitalism based on investments and returns in these new bio-industries: after mercantile, industrial, and knowledge capitalisms... based on a self-organizing and self-replicating code that harnesses both the results of the information and new biology revolutions and brings them together in a powerful alliance that enhances and strengthens or reinforces each other.
>
> *(Peters, 2012, p. 105)*

This is not ultimately about digitizing biology but is about the application of biology to information. Peters suggests that the future will be based on the 'biologisation' of the computer and its self-replicating memory. In practice, bio-informational capitalism has been applied and has developed new biology applicable to informatics in order to create new organic forms of computing and self-reproducing memory that in turn has become the basis of bioinformatics. As Peters so aptly remarks, one of the key questions is about who owns what, and most poignantly, whether synthetic biology could lead to claims to intellectual property that involve DNA and genome codings and the patenting of new life forms (p. 109).

Surveillance capitalism

Since the other forms of capitalism, a new form has emerged, that of surveillance capitalism.

Zuboff (2019) argues that we have already entered a new and unprecedented era, that of surveillance capitalism, in which the dominance of the main technology companies, notably Google, Facebook, Apple and Amazon, have adapted capitalism to suit their own ends and over which the rest of us appear to have little or no control. This form of capitalism is a new logic of accumulation, based upon the realisation that the apparently useless extra data that are now available through the mechanisms by which the digital can track our movements, desires and behaviours, in turn, produces a behavioural surplus which is then turned into prediction products. The technology companies extract these data without our consent or consultation, so we become the raw material and not even the product of this process. By this means companies can reduce the levels of uncertainty about our future preferences, be it those to do with consumption or politics, and thus shape our behaviour towards guaranteed outcomes. Even though we may be aware of this, most of us seem content to trade off our privacy and the intrusion of such targeting for the supposed benefits of convenience and ease of communication.

Cyber capitalism

We define cyber capitalism as the specific use of the internet to acquire funds for commercial gains. An example is commercial companies using digital mourning labour (the commodification of death for profit) to sell their products by capitalising on the grief and mass mourning following the death of celebrities. For example, after the death of David Bowie in 2016, the music and fashion industries both shared their grief on social media using images such as the thunderbolt, the signature sign of Bowie. This cyber capitalism therefore raises questions about the ethics and values of the digital world (Table 1.2).

TABLE 1.2

Capitalisms and AI

Term	Definition	Example	Impact of AI
Techno-capitalism	This is the commodification of knowledge resulting in the corporatisation of inventions and innovation.	Technology is used as the security interest for securing financing.	The commodification of knowledge through surveillance and LLMs will result in an increase in techno-capitalism.
Cognitive capitalism	The accumulation of immaterial assets (such as intellectual property rights) and the brandings and patenting of such assets.	Microsoft's putting a patent on the scheduling of meetings resulting in a licence fee on Android phones.	The growth and developing of LLMs will mean that people's data will continue to be stolen and reused.
Bio-informational capitalism	The application of biology to information.	A possible future where what occurs is the 'biologisation' of the computer and its self-replicating memory.	The growth in Deep Learning is likely to affect this as a possibility.
Surveillance capitalism	The commodification and selling of personal data by corporations.	Collection of data such as photos, social media posts, physical locations and product keywords which is sold without realisation or permission.	The growth of semantic AI will increase the types and levels of surveillance possible.
Cyber capitalism	The use of the internet to acquire funds for commercial gain.	Phishing and scams as well as hyper-personalised advertising.	Semantic AI will increase phishing and scams.

Conclusion

This chapter has provided an overview of different forms of AI in order to set the scene for the ethical concerns that will be discussed in the rest of this book. It has also raised the issue of capitalism and defined some of the many new formulations of this. These new forms of capitalism are important because of the way they affect and are affected by AI. In a litigious world where greed is increasingly central, it is important to understand how we and our assets are being managed often unknowingly. The challenge about the realistic and ethical use of AI is managing it in the face of rapid and often uncontrollable change. In Chapter 2, we examine the realistic and ethical use of AI in more detail, from analysing Big Data and data analytics to an in-depth ethical analysis of Neural Networks, Deep Learning and LLMs.

2

Understanding AI

Introduction

This chapter discusses AI in a more in-depth way than Chapter 1, with the first section explaining the specifics of algorithms and Big Data and their various implementations and challenges. The second section explains the relationship between AI, Machine Learning, Neural Networks and Deep Learning and in particular how Neural Networks become Deep Learning. The final section presents how an AI model is built as and explores some of the problems that emerge during AI development. The overarching argument is that whilst AI may be seen as some extraordinary innovation, numerous difficulties still remain and therefore understanding its complexities is paramount.

Analytics and Big Data

The following section presents the more traditional forms of AI and how these relate to analytics and data analysis and then discusses the role of Big Data as it applies to AI. The last part of the section analyses the workings behind the algorithms that govern internet searches, shopping systems, and social media platforms.

Analytics

Analytics is defined here as the computational analysis of statistics or data that locates meaningful patterns that then may be used in decision-making. Whilst in business intelligence and analytics there have been attempts to classify analytics research into a clear typology, Table 2.1 illustrates that across the field of business and higher education the issues are murkier and more complex. Thus, it is possible to see multiple and overlapping types: namely, (big) data analytics, text analytics, learning analytics, academic analytics and others such as network analytics and mobile analytics. Examples of some of these are illustrated in Table 2.1 below.

DOI: 10.1201/9781003622949-3

TABLE 2.1

Assorted analytics (Savin-Baden, 2024)

Form of analytics	Context	Purpose	Related research
(Big) Data analytics	Commercial contexts and data warehousing but also increasingly used in higher education.	Development of data mining algorithms and statistical analyses.	(Ranjan & Foropon, 2021) (Alkhalil et al., 2021)
Text analytics	Information retrieval and computational linguistics.	Discovering the main themes in data such as news analysis, opinion analysis and biomedical applications.	(Moreno & Redondo, 2016)
Learning analytics	Module/course-level Departmental level.	Analysis of student engagement, predictive modelling, patterns of success and failure.	(Dincer et al., 2019)
Academic analytics	Institutional National International.	Analysis of learner profiles, performance of academics, knowledge flow, research achievements and ranking.	(Paz & Cazella, 2019)

Big Data

Big Data, it seems, is a significant concern for many, largely because it is complex, unmanageable, and open to misuse. For many researchers across the disciplines, Big Data is expected to offer new insights into diverse areas from terrorism to climate change, whilst also being troublesome since it is perceived to invade privacy and increase control and surveillance. There are also definitions that focus on the economics of Big Data. For example, Taylor et al. cite examples such as the number of variables per observation, the number of observations, or both, given the accessibility of more and more data (Taylor et al., 2014). Conceptions of Big Data tend to fuse across the realms of both collecting large data sets and the processes of managing such data sets, as well as examining how, by whom and for whom the data sets might be used for everything from marketing to learning. Kitchin (2014) and Kitchin and McArdle (2016) delineate Big Data in terms of:

- Huge in volume, consisting of terabytes or petabytes of data.
- High in velocity, being created in or near real-time.

- Diverse in variety, being structured and unstructured in nature.
- Exhaustive in scope, striving to capture entire populations or systems (n = all).
- Fine-grained in resolution and uniquely indexical in identification such as digital CCTV and the recording of retail purchases.
- Relational in nature, containing common fields that enable the conjoining of different data sets, such as customer transactions and data uploading.
- Flexible, holding the traits of extensionality in that it is possible to add new fields easily and expand in size.

For scientists, Kitchin's stance seems a good fit, but those in social sciences and humanities tend to use the term data differently. For example, researchers in the social sciences see Big Data encompassing not just large data sets but also the complexity of how data are synthesised, the ways in which tools are used and who makes which decisions about how the possible imbalances between collection, management and synthesis of these data are managed. There have been suggestions that Big Data and analytics are necessarily objective, but the complexity of their use in different disciplines has meant that there is little unity about how these data should be analysed and used. It seems, for many researchers in this field, that the focus is on the analysis of data and the development of complex analysis procedures for Big Data, rather than asking critical questions about whether it is new and what can and cannot be done with it. The result is that across the literature there is a wide range of positive and negative claims that need to be acknowledged, namely:

Claim 1 Big Data speaks for itself. This is clearly not the case since analysis and mapping are researcher-driven. There is a need to ask not just what might be done with Big Data but why (and if) it should be used in particular ways – as well as how big and small data might be used together.

Claim 2 There are many good exemplars of Big Data Use. This is not the case, particularly in the social sciences and education, where the landscape is complex and varied. Furthermore, the archive released by Snowden (https://snowden.xsurveillance.site/) indicated that the e-mails, phone calls, text messages and social media activity of millions of people around the world had been collected and stored, and then without consent had been shared and sold.

Claim 3 There is integration and understanding across the disciplines. Whilst some universities have shared forums for Big Data, much Big Data remains in disciplinary silos. There is a need for greater interdisciplinarity and large teams to work together coherently.

Claim 4 There is a coherent view about how learning and academic analytics should be used. It is evident that institutions already seem to be finding themselves having to balance students' expectations, privacy laws, tutors' perspectives about learning and the institution's expectations about retention and attainment.

These claims and the challenges they raise exemplify the need to consider issues of plausibility and honesty in Big Data research. What is often missing from claim-making debates is how power is used, created or ignored in the management and representation of Big Data, or where voices are heard or ignored, privileged or taken for granted.

Whilst there is a tendency to believe that 'big data' might be bad and possibly dangerous, there are many types and uses. The challenge of Big Data has been, until fairly recently, portrayed as something that is straightforward, clear and easily delineated, when in fact it is none of these, and there is still relatively little consensus about how it might be defined. Zuboff argues that we have already entered a new and unprecedented era, that of surveillance capitalism, in which the dominant main technology companies have adapted this capitalism to suit their own ends, and over which the rest of us appear to have little or no control (Zuboff, 2019). This lack of control is shown in the way the internet is used today; every website clicked on presents terms and conditions, and every app downloaded wants access to the gallery or to the camera for whatever reason. The users have no idea why these apps need access to various features and also what they are collecting. Cookies, for example, collect and store relevant information from a site so it can be passed between different pages of the site and when you return to it at a later date, you can easily return to the section that you left. Whilst cookies are one way of giving the site memory, they also allow the site to collect various data on the user, their computer and so on. All it takes is one accept button to allow the presence of big data. Mike Savin offers his view from an IT business perspective.

EXPERT REFLECTION
Mike Savin, *Senior IT leader,*
National Energy System Operator, UK

AI AND FRAUD

As someone who has worked in business for over 25 years, I consider that Generative AI can be helpful in its simplest form – answering questions, finding facts and helping with suggestions. However, like any computer code used to manage a manufacturing plant, fly an aeroplane or manage a car's engine, code can be wrong, have bugs or simply be 'adjusted' for a different use or scenario. Incorrect code is more than a mistake – in AI terms, it's a blind spot to our reality. However, not all the rules of engagement – guidelines, guardrails or boundaries of use – are clearly articulated for organisations building new AI code constructs. Pattern-based rules in the software and a click-and-drop approach to technology are now able to manipulate mankind. This is true of social media in influencing people or people carrying out business, as shown in the following example:

> A finance worker at a top bank paid out over $20 million USD to fraudsters using deepfake. The video call was an orchestrated video team meeting where IT software had been used to pose as the company's chief financial officer. The elaborate video call included other members of staff, all of which were formed from deepfake software. The banker believed in the video call and agreed to move 25.5 million US dollars as directed by the deepfake Chief Financial Officer. The banker put aside his concerns after the video call because other people in attendance appeared like colleagues he knew.

The case is one of several recent episodes where fraudsters have used deepfake technology to modify video and voice-overs to cheat people out of money. The scam was only stopped when the banker unilaterally telephoned the UK head office to check on the details of the call and the attendees.

A further concern is broken code. Code breaking has always been with us. Whether it be picking a lock, opening a safe, or cracking the Enigma code at Bletchley Park to track German submarines in the North Atlantic. Code-breaking Generative AI seems more challenging, and it is. However, today, we still have code breakers that can help. Linguistics can pick up inclinations in voices, the shady areas of narratives and indications of whether the voice or the video is real or a deepfake. Leading universities in the UK and elsewhere are researching the skills needed to address Deepfake programmatic AI tools. The tipping point may come when the Generative AI algorithms become so advanced and unregulated that they are able to invent without rule and regulation and so create pseudo-organic code on binary IT systems, or when organic AI becomes a reality to humans. Organic AI computing will have a highly significant impact on our race. If technologies in the next 50–100 years are left unchecked, they will become unethical to how we see life today, as machines will be able to invent their own machines and organise them to do work, communicate and run teams or even organisations.

What future lies ahead depends on how ethics, rules, family and belief systems are built and embedded into new super-advanced non-generative AI organic machines. Whether in a global or protectionist world, the AI age is here and will need a revolution in both ethical and cyber trade laws, together with strong rules of engagement to ensure humans can co-exist with technology through the 21st, 22nd and into the 25th century.

This reflection illustrates the troublesomeness of our situation and introduces questions about whether we will still be on this planet in the 25th century, questions that are raised too in Chapter 5, Climate Change.

When Big Data gets too big

The problem with aggressive data collection is that too much data are collected, leading to huge datasets that then cannot be sifted through effectively to find the correct trends. The two main problems with huge datasets are storage due to their large size and the model-building time due to each model having to analyse such a huge dataset. Furthermore, with so much data, there is the problem of the AI making mistakes due to problematic data, which is more commonly known as 'noise' or 'noisy data'. Noise is defined as random and/or irrelevant data that can cause inconsistencies and various problems within the AI. These data can be outliers (data points that differ significantly from other observations), incorrect data or mislabelled data points, which all lead to the AI making the wrong conclusions and incorrectly mapping trends, which then results in a significant level of inaccuracy. This, coupled with some users' growing awareness of the problem of data collection as well as an increasing number of tools to assist them, means there is an increased amount of noise being generated. The tools can take the form of adblockers as well as browser add-ons such as Do Not Track Me and Ghostery, which provide cookies with dummy data. This dummy data significantly worsens the problem of noise and therefore creates more problems within big data.

Aggressive data collection also extends to seemingly innocuous website features, such as the 'I am not a robot' – these Captchas are based on analysing your browser history, cookies, mouse movement, and several other factors to determine whether or not you are a bot. This data collection from Captchas has been criticised due to shortcomings, such as its potential susceptibility to brute force attacks. Yet the tools discussed often enable Captcha to identify bots which do show their effectiveness. Furthermore, this is a double-edged sword, as the image Captcha is then triggered more often, and this image identification is then used to train Waymo's self-driving car systems (Fang, 2022). This shows that there is no easy solution to the big data collection issues despite the many attempts at mitigation. Furthermore, given the amount of image identification used by the Captcha systems, it will be curious to see how effective the Waymo AI system is as it is deployed. This is due to it being trained by humans who often make mistakes when completing the Captcha. However, according to Waymo, their self-driving car is already safer than humans by a significant margin (Waymo, 2024).

Whilst harnessing the general populace to inadvertently help with the training of AI and management of Big Data in this way is one approach to mitigating the problem, it is far from the perfect solution. Yet the problem of Big Data is not likely to disappear due to companies continually hoarding data without necessarily knowing what to do with all of it. Therefore, there needs to be a better approach to data collection, as well as better tools for users to protect their own data.

Algorithms

Algorithms are a series of instructions or procedures that collect and encode data and, based on specific calculations, transform data into output that brings about a desired outcome. In daily life, algorithms affect media recommendations people receive, such as reviews of books and equipment and Twitter feeds, as well as more personal areas such as Google and Fitbit tracking. Algorithms can be either simple or complex:

- Simple algorithms are used for almost all computer programmes, even Complex Enterprise Resource Planning and Customer Relationship Management Systems, since they are highly linear and predictable.
- Complex algorithms are programmes such as Machine Learning, Deep Learning, Neural Networks, Bayesian networks and fuzzy logic where the complexity of the inner code starts to move beyond simple linear relationships. Many systems currently referred to as AI are complex algorithms.

(Abbreviated from Burden & Savin-Baden, 2019, p. 27)

Nevertheless, it is important to remember that algorithms are created by humans and that the reproduction of world views and stereotypes is, as it were, 'designed in' by humans. A further concern is assumptions that are made by humans about the impact and control algorithms may have on our lives.

However, the major concern with algorithms is their modus operandi. For those who are often on TikTok and YouTube as well as other social media, there is always talk of 'The Algorithm'. It is portrayed as this mysterious ungoverned force that can hand out fame, and there is an ever-growing armada of videos discussing the ways of how to 'beat' or 'trick' the algorithm. The issue with the algorithm is that no one really knows how it even works, which is as ludicrous as it is true. This is due to most algorithms being black-boxes which investigate the private data of the users whilst humans cannot do so due to data privacy laws. Therefore, only the result of the algorithm can be viewed, not how the algorithm came to that result.

There are some elements that are predictable, for example, a social media algorithm is tailored and targeted to each person, so if a user looks at content featuring dinosaurs, they will be shown more dinosaur content and will also be directed to related content within that genre. Although this makes sense, the recommendations to related content can often be problematic or repetitive; sometimes, it only takes one click to trigger the algorithm into bombarding the user with this type of content.

When considering the unknown elements of the algorithms, these algorithms are personal to each user, and the algorithms learn from the private user data. Thus, it learns from the preferences of that user, such as what they interact with and what they 'like', which is kept private from social media companies for data protection. This is why social media corporations cannot explain why users are recommended problematic or explicit content because they cannot see what the user looks at, and at some level, it is the fault of the user for interacting with a potentially controversial topic. These companies are reliant on the algorithm to tell them what is trending and what content is likely to set a trend or go viral, yet even if that content is produced, there is no guarantee of it whatsoever. Although social media companies can add new features to their social media platforms and tune the algorithm to boost certain topics and push certain types of content, this can often have unintended effects. For example, if a video is watched at half speed on YouTube, then the watch time doubles: 2 minutes becomes 4 minutes. This then boosts the video's rating, as the YouTube algorithm considers watch time as a metric to gauge whether a video is good or not (The Spiffing Brit, 2024). This is caused by YouTube adding a speed option and not considering the consequences. Whilst this activity is not used much due to poor quality sound, it is worth considering due to the design faults it indicates, since when developing this feature, the effect of watch time was not considered. Furthermore, the user base of the social media platforms can shift the algorithm vastly depending on their opinions and the content to post. One such example is when YouTube Shorts was released. Due to the popularity of TikTok and its short-form content, the popularity of YouTube Shorts exploded and now makes up a significant portion of the content on YouTube. So, the unpredictable nature of the algorithms that govern us day to day is partially unpredictable because of human nature.

Understanding Neural Networks and Deep Learning

As mentioned in Chapter 1, Neural Networks are advanced Machine Learning models that simulate the human brain's functions. Deep Learning utilises Neural Networks using many hidden layers – hence it is referred to as deep and requires large volumes of labelled data and processing power. This section provides a more in-depth explanation of these terms in the context of AI and Machine Learning. The relationships between them are represented in Figure 2.1 below.

The diagram illustrates that machine learning is what is being used day to day, such as in social media algorithms, Siri, autocorrect and predictive text. This is the AI that has been in use for decades. Neural Networks and Deep Learning are more advanced and accurate versions of AI which have developed rapidly in recent years. To recap:

1) Artificial Intelligence – this is the idea of computers being able to solve tasks that normally require a human level of intelligence; this includes:
   - Basic/procedural programming – building a programme with a set of steps that are followed, often with multiple paths for multiple outcomes to accommodate most scenarios. One such example

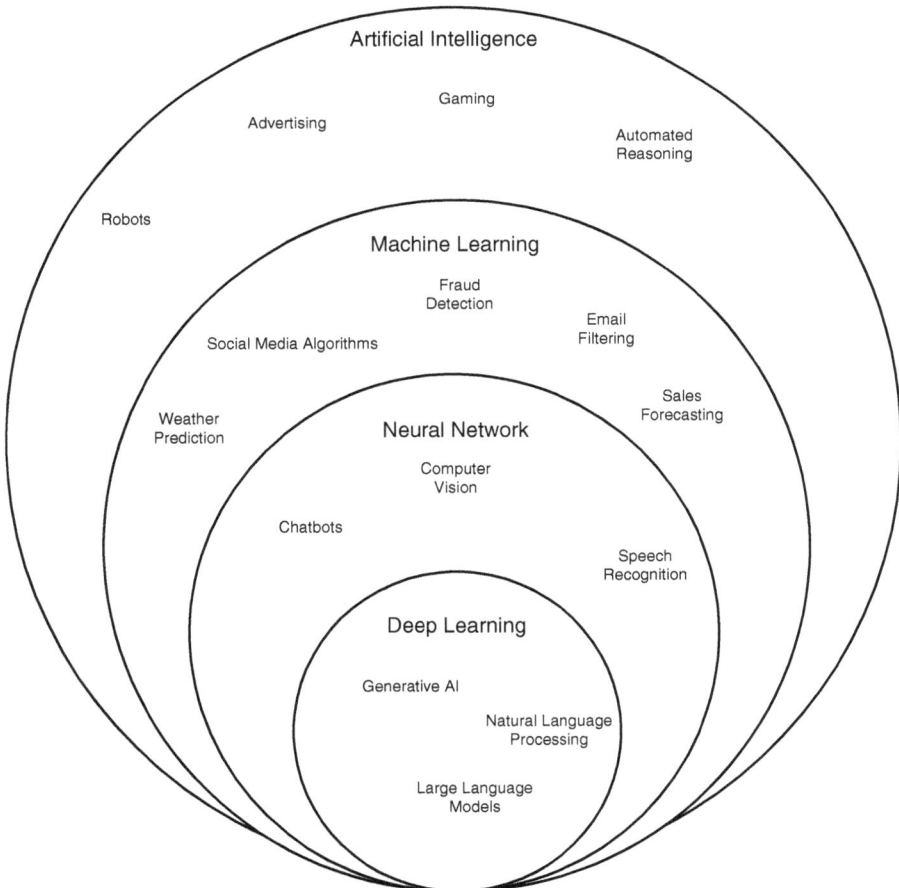

FIGURE 2.1
Diagram of the relationship between AI, Machine Learning and Deep Learning.

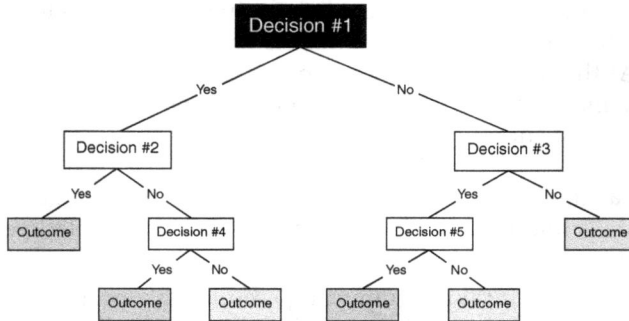

FIGURE 2.2
Decision tree diagram.

is NPC (non-player-characters) in gaming or simple chatbots which often use basic decision trees (Figure 2.2) for their dialogue and actions.

- Simple algorithms – These often take the form of search algorithms used for search engines and databases as well as optimisation and sometimes for gaming for searching environments as well as pathfinding, i.e. the way a robotic cleaner works out how to move around a room without hitting anything, then how it calculates the optimal cleaning routine once a safe route is established.

2) Machine learning – this is a branch of AI that focuses on machines imitating the way that humans learn and improve performance and accuracy by exposure to data. Essentially the common approaches are providing a dataset to an algorithm and asking it to predict an outcome based on common features. These are just a few examples of the techniques used.

- Random Forest – This model takes random parts of the dataset, runs each part through a decision tree and then averages the decision tree to draw a conclusion. This means that it can draw a variety of conclusions and take the best of all of them.

- Naïve Bayes – This model assumes that the features of the dataset do not affect each other and treats all data as equal. The issue that arises with treating all data as equal is that some data is more important, and some of the data points do affect each other. For example, a dataset containing weather information would commonly show colder temperatures when it is snowing.

- K-Means Clustering – This model works by creating central points known as centroids and grouping similar data around these points to create clusters. The centroids are updated once all the data is grouped. When new data is added for prediction, then it is based off the centroids. However, if the clusters overlap due to the data having similar features, then it means that the model can group data incorrectly.

3) Neural Networks – A subset of Machine Learning that makes decisions in a similar way to the human brain by using a network of neurons in a similar manner to biological neurons.

4) Deep learning – A subset of Machine Learning that uses multiple layers of a Neural Network.

- Convolutional Neural Networks – works on grid-based data and filtering to draw conclusions. For example, if the network is passed an image, then multiple filters are applied to recognise specific patterns within the image, such as curves and straight lines. The model then learns the relationships between these filters before extracting the most significant features and learning from them.

- Recurrent Neural Networks (RNNs) – works on sequential data and uses loops to have a memory. For example, if the network is given text in which there is the phrase 'Apple is red', then next time it detects the word 'apple', it will have a higher probability of putting the word 'is' after it.

- Long Short-Term Memory (LSTM) – a type of RNN that controls what it remembers and what is not, instead of trying to remember everything. This is done to build a longer-term short memory by only remembering important parts, i.e. the features it considers most relevant to the output.

Conventional Machine Learning versus Deep Learning

At a basic level, Machine Learning is similar to drawing a line of best fit between the points on a graph. Various methods such as Random Forest, K Nearest Neighbour and Gaussian Process are just different approaches applied to drawing the best-fit line. Figure 2.3 illustrates how various methods arrive at different 'best-fit' lines. For example, linear regression's best-fit line is always straight, whilst in the Gaussian process, the line is straight with some peaks and troughs to try and encompass points. Random forest seeks to encompass most of the points, and so its line is quite irregular, and so it is likely to overfit due to containing so many of the points. In this example, the neural network draws the most suitable best-fit line that bends and shakes around the points without overfitting (see Figure 2.3).

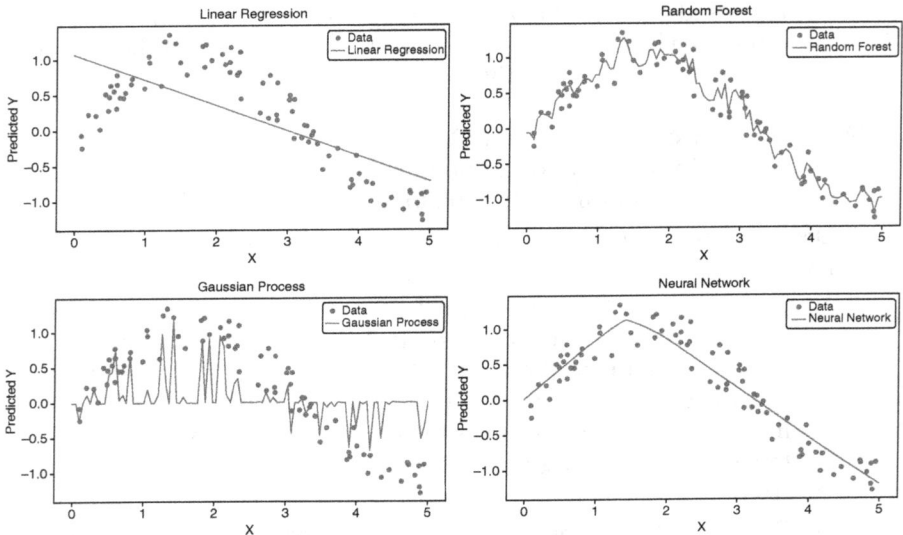

FIGURE 2.3
Comparison of AI models drawing the line of best fit.

A Neural Network is considered deep if it has three or more layers to it. Each layer works on processing and/or transforming the data before passing it to the next layer. Hence, any Neural Network deeper than three is considered deep learning. Deep learning can be applied to practically every problem that falls within AI, although it is far from foolproof. The main problem with Neural Networks is that they require an enormous amount of data in order to run and make accurate conclusions, even more so than conventional AI, meaning that they cannot be applied to very specific situations, only more general problems. Furthermore, due to the high amount of data required and the specific learning needs of a Neural Network, it demands significantly more computational resources than a conventional AI approach such as K Nearest Neighbours. Another problem is that Neural Networks work by winding their best-fit lines around the points and become highly sensitive to these points, which means there is a risk of overfitting (see Model building). The network widely changes if these points/inputs are changed, meaning that there is an increased risk of it being fooled or hacked (Yuan et al., 2018). Therefore, the unpredictability of neural networks can make them unsuitable for applications where safety is paramount. A good example of this is if a Neural Network is trained on a dataset of car images produced by a photographer who has added their own watermark in the corner of each image. If the network is then provided a photo of the car without the watermark, then it may have trouble identifying the ca due to it being trained on images with a watermark. This is because the network identifies it as a relevant part of the image (Lapuschkin et al., 2019). Whilst this error is quite easy to explain, most unexpected outputs from Neural Networks are not. This is due to the black box nature of Neural Nets, i.e. only the input and output can be viewed and the internal process of how it reached that output conclusion cannot. This means that they have low interpretability and explainability which makes debugging and analysis extremely difficult (Fan et al., 2020).

Furthermore, Neural Networks have significant problems with long-term memory and often forget key parts of information when presented with brand-new information. This is known as catastrophic interference or catastrophic forgetting (McCloskey & Cohen, 1989). This issue is why ChatGPT sometimes loses context of the situations the user provides and incidentally why some AI systems have a specified memory limit. This memory limit does place a significant hindrance on the applicability of Neural Networks and means updating Neural Networks to always hold the most relevant information can cause significant problems. Yet there are various mitigations and ways to address these, such as Bayesian Neural Networks, but it is far from a solved problem. So, whilst Deep Learning and Neural Networks are often portrayed as a valid solution for a significant number of problems, they are far from a miracle solution.

Model building

The typical process for model building begins with locating a dataset or collecting the data yourself. Then the data is divided, typically 70%–30% into a training set and a test set. Sometimes, an additional test set known as a validation set is also made, with a typical split being 70%, 20% and 10%. This is then followed by exploratory data analysis (EDA) which is used to identify patterns within the data as well as map problems and spot errors. For example, if there are points which show a high correlation merely through coincidence, then sometimes, these are removed to prevent the model from analysing that particular pattern. The problems identified in the EDA should then be removed in the next stage, which is data pre-processing. It is important to note that EDA and data pre-processing are only performed on the training set to avoid data leakage. Data leakage is when elements of the training set are mixed with the test set, causing a bias or unnatural increased accuracy in the test set, almost like peeking at the answers on a test. However, a few choice elements of the data pre-processing stage may be performed on the test set, where it can be ensured that it will not affect results, such as dropping unnecessary columns and renaming features. The next step is model building or model development where the AI model is run on the training set. Then the model is evaluated using the test set and/or the validation set. Then the model is tuned to improve the accuracy as much as possible by using other evaluation metrics such as F-1 score as well as false positives and false negatives. However, a model that has 100% accuracy is not as perfect as it seems.

There is a saying in Computer Science: 'the model is only as good as the data'. This becomes particularly key when considering the data used. As discussed earlier in this chapter, big data is a growing concern across various disciplines and will continue to be a continual problem. There is the inherent problem of poor data quality, which can adversely affect AI models such as by causing bias. This calls to mind another phrase, 'data doesn't lie but the data scientists do'. This is generally due to data scientists cherry-picking data through either removing outliers from the dataset or only selecting the dataset entries that fit with a specific conclusion. This can lead to model accuracy increasing dramatically. The removal of outliers and anomalies is relatively common in model building, since some of these are genuine mistakes or impossibilities such as a woman who is eighty years old and pregnancy being present in a medical dataset. Whilst it makes sense to remove such obvious mistakes, other records within a dataset classed as anomalous or outliers should probably be retained to ensure that the model can perform on edge cases or specific cases that are uncommon. However, data scientists are often pushed to attain high accuracy by stakeholders. Yet if the model cannot predict certain specific cases or edge cases, then this can become an issue, especially if a medical model is unable to identify a rare illness due to misdiagnosing the presentation of an unusual set of symptoms that are nevertheless consistent with the rare illness.

The high-accuracy models that corporations like to promote in press releases and within news articles indicate an additional deeper story. If an AI model has a high accuracy of 90–100%, then there is likely to be overfitting within the model. Overfitting, a similar idea to favouritism, means that the best -fit line is essentially too good. This becomes a problem when an overfitted model is presented with new data since it is unable to assimilate the new data into the trend it mapped with the previous set because that was too specific. Therefore, it leads to a poorer model overall. High accuracy counts should be treated with more suspicion than praise. Furthermore, biases are often published on company press releases, and therefore, it is also worth considering what biases are likely to be present within models, such as gender and racial bias. Image generators tend to have a bias towards white people and have a habit of amplifying stereotypes. When Gemini launched with the intention of providing racially diverse images it portrayed Nazi soldiers from 1943 as People of Colour (Robertson, 2024). This happened because Google set up Gemini in an attempt to fix the general bias towards white people and inadvertently went too far the other way.

This bias and problematic data lead to overconfidence and hallucinations, as well as outright mistakes by the model. ChatGPT for example is more likely to make something up rather than say, 'I don't know.' However, the greatest flaw with current AIs is the fact that people put them on pedestals, assigning them a humanity that they do not possess, placing faith in them that they do not deserve, going so far as to trust technology with significant portions of their lives. This is most prominent with Large Language Models and their mass deployment, which is covered in the next chapter.

The next chapter also discusses the unethical nature of utilising LLMs using datasets containing copyrighted material. To preface this, all the model techniques and methods shown above also require datasets to run. Yet these datasets are commonly created by research groups, companies, and individuals, generally by collecting the data themselves or by collecting it from somewhere else and asking permission, as well as often adding and/or updating data. These sets are then often provided for free on sites like Kaggle and UCI for anyone to use. The main difference is that these datasets have been created ethically, whereas the datasets used for ChatGPT contain copyrighted material, as ChatGPT has been trained on books, artworks and other media without permissions or licences. The full extent of the unseen theft of works is covered in the next chapter.

Conclusion

To conclude, this chapter has presented the basics of AI and covers some of the unique problems with the various systems, as well as the issues that stem from humanity itself. Artificial intelligence is a tool with untold possibilities yet is far from the miracle solution to everything that some describe, whether it be simple algorithms or advanced LLMs which have a great deal of complexity behind them and often seem inaccessible to the general public. Furthermore, these AI systems often provide disingenuous results that are difficult to explain or are simply unexplainable. This leads to many mistakes and a lot of unpredictability within the AI world. Therefore, the challenge is to change the perception of AI by humanity through education, as few are aware of or know the vast array of problems as well as the ethical consequences behind them. These consequences are covered in the next chapter, which discusses many of the uses, dangers, and horrible consequences of AI. Yet solutions are presented and some hope offered.

3

Large Language Models and Generative AI

Introduction

This chapter begins by examining how Generative AI works, exploring the Turing test and how Large Language Models are changing and being developed. The second section of this chapter analyses some of the facts and myths related to Large Language Models in terms of whether they do 'understand' or only function on the basis of statistical occurrence models. It then explores understandings of trust in relation to Large Language Models, particularly ChatGPT, as well as the extent to which people do in fact trust them. The final section examines theft and in particular the unseen theft that is occurring without permissions, particularly for those that have created art and literature.

How does Generative AI work?

The types of Generative AI and LLMs that are currently at large in the modern world range from AI image generators to deepfakes and 3D landscape generation. Yet these can be simplified into text generators and image generators. The idea behind a text generator is very straightforward. The model is provided with a huge amount of text data so that it can learn the patterns of the language, with the aim being that it can predict the next word in any sentence. LLM dialogue functions predominantly on educated guesswork, which means there is no functional intelligence or sentience behind it. It simply says there is a 70% chance of the next word being 'X' so it will output X. However, this means that text generators do come with some inherent problems such as faulty guesswork rather than admitting defeat and making no prediction at all. Since there are little to no fact-checking systems implemented, this becomes 'artificial ineptitude' whilst users equate 'intelligence' with what is essentially a glorified guesswork machine. The idea behind image generators is much the same; however, it is image data that the AI

DOI: 10.1201/9781003622949-4

learns from, and when generating images, it first analyses the text the user gives, then relates those to images associated with that text, and then does its best to recreate elements of those images. The reason why both are referred to as Large Language Models is that a text generator is a text-to-text generator, whilst an image generator is a text-to-image generator. It is interesting to consider what people believe and understand about LLMs, since a considerable amount of misinformation is being endorsed across the internet by both people and algorithms. First, we examine some of the underlying issues such as the Turing test and early work on language simulation, before examining some of the facts and myths.

The Turing test

The Turing test was first proposed by Alan Turing (1950) and was meant as a test for machine intelligence based on whether a user sitting at two computer terminals could tell on which one they were conversing with a human and on which with a computer. Since 1950, variants of the Turing test have been instrumental in the development of today's computer-based AIs. The test has been much discussed ever since (for example, Epstein et al., 2009; Saygin et al., 2000). Whilst it is no longer seen as a test of 'intelligence', it is certainly a useful, if not ideal, test of chatbots and natural language capability. A system which is able to pass such a Turing test in a particular setting is often described as being 'Turing-capable'. Turing himself allowed only digital computers, arguing 'This restriction appears at first sight to be a very drastic one. I shall attempt to show that it is not so in reality' (Turing, 1950, p. 436).

Every few years, there is a newspaper headline claiming a new machine has passed the Turing test, but these claims are usually immediately disputed. Annual applications can be made for the Loebner Prizes which appraises attempts for the successful completion of the Turing test. The prizes differentiate between a Bronze award for best chatbot that still fails to convince any judges of its humanness, a Silver award for a chatbot which passes the Turing test in text-chat only, and a Gold award for a chatbot which passes the Turing test in speech and video. It has been suggested by many researchers (see, for example, Burden, et al, 2016; Burden, 2020) that the Turing test is too narrow and there is no scope to add in additional options. Since these studies there have been many developments in chatbots, LLMs and the simulation of human conversation. However, there are now arguments for a need to shift away from mimicking in LLMs and move towards understanding. However, Grzankowski (2025) stated that whilst the following features of intelligence have been suggested, they do not really guide us effectively when seeking to benchmark intelligence. Features of genuine intelligence comprise

- generality: principles applicable across diverse contexts
- flexibility: solves novel problems
- adaptivity: learns and refines from feedback
- goal directedness: learning works towards a goal

Whilst he suggests these features are a good starting point, it is clear that many LLMs can do a lot of this essentially, they do all the 'right' performance. However, he suggests these are not really properties about performance but about underlying capacities.

What is evident is that the rapid changes in technology and in particular the rise in the use of AI have resulted in confusion about some of the 'truths' associated with these technologies.

Facts and myths

This section begins by examining the facts and myths of LLMs in a way that demonstrates that decisions about what is true and false are complex and not particularly straightforward. For example, it was thought for some time that LLMs only functioned on statistical occurrence, and recently, this assumption has been challenged, as will be discussed below.

Simulation of human conversation is new

With the growth and popularity of apps such as Siri, many people have assumed that this technology was developed recently. In fact Block (1981) presented an argument to counter the Turing test as a test of intelligence and undertook a thought experiment. In the computer system (now commonly referred to as Blockhead), he argued that a dialogue can last any duration but that there are constraints on the number of legitimate replies that can follow the initial sentence. Block argued that the internal mechanism of a system is important in determining whether that system is intelligent and claims that he can show that a non-intelligent system could pass the Turing test. What is important about this work is that whilst the computer can simulate human conversation perfectly by making assessments from a huge conversation tree and then responding, it still lacks understanding. However, many researchers have questioned Block's arguments on the basis that the Turing test is an assessment of the interaction of the entity being measured within a specific environment and time, and yet Block's experiment can take any duration whatsoever. What is clear is that work on human simulation, of whatever sort, is not new but has been developed over many decades.

The major limitation of the Turing test is that it does not factor in how the intelligence of the computer system was developed in terms of how the computer system was designed to come to the conclusion of what dialogue output would be rendered. It focuses only on the dialogue itself. One perspective that can be taken is that the Turing test has not really been designed as a test for its AI system but for its humanity. With this view, the test only focuses on how low the requirements have to be for a human to consider an LLM to be human or to have some level of humanity. Furthermore, some users of LLMs like ChatGPT use please and thank you when communicating with the LLM when there is functionally no need to do so. The fact they put some element of politeness into their prompts does demonstrate some level of connection (Yuan et al., 2024). The fact that some users put significant faith and connection in the LLMs demonstrates that these users attribute intelligence and humanity to something that fundamentally lacks it. They trust LLMs to answer texts and emails at a human level of competence and intelligence. The question of whether we are communicating with an AI or human is something we should ask more and more as LLMs become increasingly integrated throughout society. Incidentally, most of the YouTube support system has become automated by ChatGPT-enhanced emails. In practice this means that the user risks being obstructed by an AI that simply recites what it says in the terms and conditions, meaning a significant impact on the effectiveness of the support system. This was revealed when the news account Sir Swag was taken over by Russian hackers. YouTube refused to store the account, as their identity could not be proven despite them providing sufficient evidence. This evidence was suitable for multiple companies and government departments. The YouTube support emails simply copied the terms of what they needed to provide or urged them to contact their channel's agent, who did not exist. This constant regurgitation of the same terms over and over showed that either support was completely incompetent or fully automated by AI. The situation was eventually resolved when a YouTuber mentioned this to YouTube staff during a business meeting, which led to the channel being restored immediately. Therefore, this added to the suspicion that ChatGPT was being used for support emails. Later the team was contacted by a YouTube whistleblower. They confirmed that ChatGPT was often used for support emails, meaning that it is likely that the Sir Swag news team was communicating with an AI the entire time.

Considering how ingrained automated support is within social media, online shopping sites, shipping companies and other online services today, it is of little surprise that automated AI systems are being implemented. Yet this means that the previous problem of being blocked from good support due to poor AI understanding and limited human oversight is likely to become increasingly exacerbated and lead to an underreport of problems as users become trapped in a continuous loop of conversation with the AI. Although support chatbots are not generally human-supervised they do now, in the main, know to elevate the problem to a human operator, and the user is normally informed that they are speaking to a chatbot, thereby reducing the possibility of becoming trapped in a loop. However, with automated email systems, the user is possibly unaware of the exclusive use of AI and totally unaware of the level of human supervision, if indeed there is any.

So therefore, when returning to the Turing test, the key problem is the user's lack of awareness. Users often have no awareness as to whether they are emailing a human or an AI, as the responses are functionally indistinguishable. This is further exacerbated if humans put their own emails through ChatGPT first and then send that response. This, coupled with the level of intelligence and faith attributed to LLMs, means that the Turing test is more of a test of the user's perception rather than the intelligence of the AI.

The Turing test is too narrow

Burden (2009) notes that in the Turing test both the judges and hidden humans were aware that they were taking part in a Turing test, and so the dialogues that take place during the test are rarely normal. In the same paper, Burden described how a more level playing field could be created by conducting a hidden Turing test, where neither judge nor hidden human knew that the test was taking place, and in particular, how virtual environments could provide the ideal location for such a hidden Turing test. A study by Burden et al. (2016) sought to test Burden's questions as to what extent do the covert/overt and singleton/group options present earlier opportunities to pass the Turing test (or equivalent). Burden defined four possible Turing test situations:

- The robotar (computer-driven avatar) was undeclared and part of a group conversation (the Covert Group test).
- A robotar was declared present (but unidentified) as part of a group conversation (the Overt Group test).
- The robotar was undeclared and part of a set of one-on-one conversations (the Covert Singleton test).
- A robotar was declared present (but unidentified) as part of a set of one-on-one conversations (the Overt Singleton test, the original Turing test and the Imitation Game (Turing, 1950) as typically implemented by competitions such as the Loebner Prize) (Bradeško & Mladenić, 2012).

Burden, working with the University of Worcester, conducted an online problem-based learning experiment (problem-based learning is an approach to learning where students work in small teams to solve or manage a problem scenario, guided, rather than taught, by a tutor) to assess the capability and value of chatbots (both covert and overt) in an educational context. The Worcester Phase 1 experiment (Group Covert) achieved a 100% deception rate, although the Phase 2 experiment (Group Overt) achieved only a 6% deception rate (although the reticence of most students to identify the chatbot during the covert session, and their comments about its improved quality, suggest that a high level of doubt may have existed). The results would therefore appear to bear out the original hypothesis that the Singleton Covert and Group Covert conditions can indeed provide an earlier opportunity to pass the Turing test than the more traditional Singleton Overt case.

LLMs only function on statistical occurrence

Large Language Models have been trained on what is referred to as statistics of occurrence, which means that they are trained on text (words and phrases), thereby producing a model that is based on statistical information that is expected to result in similar responses to that of a human. There have recently been suggestions that LLMs such as ChatGPT and DeepSeek have semantic understanding (the ability to understand words and phrases in context along with underlying meaning and intention, but these are largely suggestions and not based on empirical research, for example, see Delcid (2022) and Tiku (2022)). Titus (2024) argues against this stance, suggesting that LLMs do not produce meaningful text but just reflect strong correlations between words and meanings.

LLMs understand

Many people believe that when they are speaking to an LLM and when asking for guidance, they make the assumption that it actually understands, and the use of kind and supportive voice fonts helps to maintain this myth. Bender and Koller provide a useful challenge through this example:

> Say that A and B, both fluent speakers of English, are independently stranded on two uninhabited islands. They soon discover that previous visitors to these islands have left behind telegraphs and that they can communicate with each other via an underwater cable. A and B start happily typing messages to each other. Meanwhile, O, a hyper-intelligent deep-sea octopus who is unable to visit or observe the two islands, discovers a way to tap into the underwater cable and listen in on A and B's conversations. O knows nothing about English initially but is very good at detecting statistical patterns. Over time, O learns to predict with great accuracy how B will respond to each of A's utterances. O also observes that certain words tend to occur in similar contexts and perhaps learns to generalize across lexical patterns by hypothesizing that they can be used somewhat interchangeably. Nonetheless, O has never observed these objects and thus would not be able to pick out the referent of a word when presented with a set of (physical) alternatives. At some point, O starts feeling lonely. He cuts the underwater cable and inserts himself into the conversation, by pretending to be B and replying to A's messages. Can O successfully pose as B without making A suspicious?... The extent to which O can fool A depends on the task—that is, on what A is trying to talk about. A and B have spent a lot of time exchanging trivial notes about their daily lives to make the long island evenings more

enjoyable. It seems possible that O would be able to produce new sentences of the kind B used to produce; essentially acting as a chatbot... Now say that A has invented a new device, say a coconut catapult. She excitedly sends detailed instructions on building a coconut catapult to B, and asks about B's experiences and suggestions for improvements. Even if O had a way of constructing the catapult underwater, he does not know what words such as rope and coconut refer to, and thus can't physically reproduce the experiment... So O decides to simply say "Cool idea, great job!", because B said that a lot when A talked about ropes and nails. It is absolutely conceivable that A accepts this reply as meaningful—but only because A does all the work in attributing meaning to O's response. It is not because O understood the meaning of A's instructions or even his own reply.

(Bender & Koller, 2020, pp. 5188–5189)

What we are seeing here is not actually understanding or intelligence per se but recognition of the signals and a highly effective predictor of the next steps in the sequence. Thus, LLMs do not learn meaning (semantic understanding) but acquire some kind of reflection of meaning that is functional but not entirely complete.

Sentience is possible

D'Silva and Turner define sentient creatures as 'those who have feelings, both physical and emotional, and whose feelings matter to them' (D'Silva & Turner, 2012, p. xxiii). Interestingly, given the example above, there appears to be more discussion in the literature about sentience applied to animals than to humans and popular culture often sees marine creatures such as octopuses as sentient.

Sentience is something more than intelligence and is certainly beyond what all (or almost all) animals show, and it is more than emotion and empathy. Sentience is in fact about self-awareness, self-actualisation and having a consistent internal narrative, internal dialogue and self-reflection. Frude (Frude & Jandric, 2015) argues that the real shift towards sentience in AI will occur when interactions between humans and AI are cumulative. This will result in long-term relationships with the person trusting the machine and in turn the machine adapting to appreciate the person's preferences. However, it is important to note that media and press reports still tend to imply that AI is more developed than it actually is, and it is important to be able to distinguish between clever algorithms, some form of well-developed virtual human and the possibility of creating a virtual persona with sentience some time before 2065. In order to create a virtual human with some form of sentience, then the ability for self-awareness and meta-cognition will be vital. There is still division amongst researchers about differences between human and animal sentience and self-awareness, which raises interesting questions about whether there could be a sentience spectrum based on what can be replicated in silicon and bits and what cannot. Thus, we suggest that sentience in AI needs to focus on flexibility rather than mimicry.

People trust ChatGPT and DeepSeek

Yu et al. (2024) argue that trust in LLMs such as ChatGPT varies culturally according to societal norms and political regulation. Table 3.1 below summarises the findings of this study.

Yu et al. (2024) also examined the stance of global IT professionals, 81% of whom used ChatGPT for learning new topics and 59% for code writing. Whilst the UK was not included in the study by Yu et al. (2024), Koopman (2025) notes that in Ernst and Young's new global AI sentiment index, 70% of UK adults used AI in their daily lives in the past six months, but only 44% used it in their professional roles.

What is notable is that the general public and a substantial number of academics use LLMs with little thought to issues of trust, reliability and ethics. We suggest there needs to be much more of a critical stance towards the use of this technology, not only in terms of trust but also in relation to climate impact, academic integrity and the unseen theft of works.

The unseen theft of works

Web scrapers such as the Common Crawl (2024) scrape the entire web for available information. Whilst this information is normally fed into search engines for indexing, scrapers like Common Crawl archive this data. Often this data is copyrighted but is collected in the name of archiving and

TABLE 3.1

Public Trust in ChatGPT (following Yu et al., 2024, with additions)

Country	Level of trust	Related work
Southeast Asia	Almost 50% indicated concerns of personal data collection, privacy and the accuracy of LLMs.	Statista (2023a)
South Korea	26.5% found ChatGPT responses to be trustworthy, 62.1% remained neutral, and around 9.7% deemed them untrustworthy.	Statista (2023b)
Japan	55.6% were somewhat interested in using ChatGPT in the future and 23.4% of participants were somewhat resistant.	Statista (2023c)
China	China has banned ChatGPT, but we suggest it will be interesting to see the progression of their own software DeepSeek.	Lian et al. (2024)
USA	The majority of US adults have not utilised ChatGPT, with only 18% reporting usage.	Park and Gelles-Watnick (2023).

preserving the data for the future. Yet it is then provided to the public freely and therefore can be used in AI datasets with limited to no consequences. Dave White offers his perspective on these issues below.

EXPERT REFLECTION

David White, *Dean of Academic Strategy (Online)*
at UAL University of the Arts, UK

ARTIFICIAL INTELLIGENCE AND THE ARTS

In addition to AI's ability to produce 'natural language' style text, there are also a plethora of platforms that can produce media such as images, video and sound. For example, a request for an image of an oil painting of a landscape produces a convincing version of a work that doesn't exist. On first impressions it appears that AI has made a successful incursion into the sacred space of creativity and the Arts, but to what extent is this the case?

The question of AI and the Arts centres on our framing of creativity and authorship. What would be of more value, an image created by a named artist or an 'identical' image created by AI? We are attracted to notions of the original, the scarce and the idea of the singular author. Once authorship becomes lost in complexity, the value of the work diminishes. The act, or the possibility, of mass production reframes what might have once been understood as creative into the mechanistic. By way of an example, I have two mugs I enjoy drinking coffee from. One is made by IKEA; it is a pleasing design and pleasant to drink from but embodies almost no cultural capital. The creative act of the original designer is disembodied-through-mass-production. The other is handmade, irreplicable and slightly inconvenient in its design. The fact that I can see the fingerprints of the ceramicist, the artist embodied through their work, is compelling and confers significant value.

However, we should not frame AI and the Arts as the artist against the machine. The Arts continue to evolve, incorporating technology into creative processes, constantly redefining and extending what we mean by creativity. The need for a 'creative', the artist, to be involved always remains. A useful allegory is the game of chess. The computational model of the game has been 'more successful' than any human since 1997, and yet the game of chess flourishes. The relationship between the digital model and the players is nuanced and has pushed the game to new heights. The technology has been incorporated into the spirit of the game itself.

The problem in this debate, as with many emerging technologies, is an overfocus on surface functionality and not the structural intent.

AI is not a technological threat to the Arts; it is a business-model threat to artists. The plundering of work to train the machine is a serious problem. It will likely lead to fewer people being able to earn a living through their creativity. We risk automating the mediocre and disassembling our creative community.

Thus, this can mean that creative works are fed into AI generators without the permission of the creators. ChatGPT for example would have learned English and good writing techniques from works such as Jane Eyre (Bronte, 1847) or A Christmas Carol (Dickens, 1843) or Lord of the Rings (Tolkien, 1955/1995). However, OpenAI would not have paid for all literature used within the training of its models, nor would it have paid for the artistic works used in the training of the art and text generators (Creamer, 2025). This is where the element of theft is present. Therefore, although the output of ChatGPT is derived from copyrighted material, the output of any LLM cannot be copyrighted since it is not written by a human. As OpenAI received significant backlash due to the theft of the artistic works, they developed a tool where artists can submit a work, and if it is within the dataset for ChatGPT, then it is removed. However, there are significant caveats to this. Firstly, the tool only allows one work to be removed at a time. Secondly, these works are only likely to be removed from the next version of ChatGPT, as to properly remove them from the current version of ChatGPT would require retraining the model. This would cost significant power consumption and money. ChatGPT is only updated by small retraining to update the current information, and data is not commonly removed in order to ensure there is no data loss from the model. So therefore this tool is relatively ineffective as a copyright mitigation. Furthermore, in an ironic twist, when the DeepSeek model was released, OpenAI suggested that they had used the same dataset and model to train it (Olcott & Criddle, 2025).

Meta and Google have also developed their own LLMs, utilising their various social media platforms such as Facebook, Instagram and YouTube. When Meta declared their intention to data scrape Instagram and Facebook, most of the European user base had the ability to opt out due to EU law. For most of the American audience this option was unavailable, leading to a significant number of users deleting their accounts and moving to a similar app called Bluesky. When considering YouTube, there is also an option where creators can allow their videos to be used for AI training. Fortunately, this is disabled by default yet there is a little tick box in the channel settings. What is even more interesting is that corporations have also sent YouTubers direct payment in exchange for being able to use their content to train AI models as shown in Figure 3.1, which depicts such an offer addressed to a Doctor Who YouTuber asking for their content to be used for training AI models.

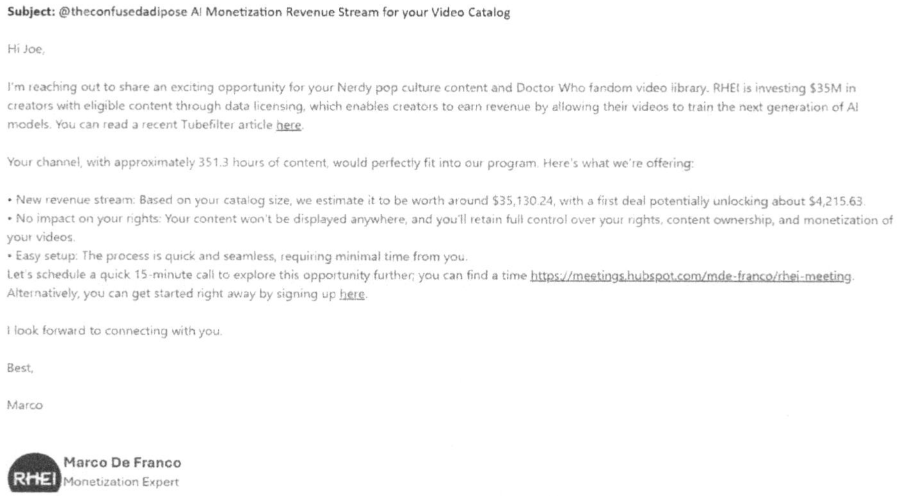

Subject: @theconfusedadipose AI Monetization Revenue Stream for your Video Catalog

Hi Joe,

I'm reaching out to share an exciting opportunity for your Nerdy pop culture content and Doctor Who fandom video library. RHEI is investing $35M in creators with eligible content through data licensing, which enables creators to earn revenue by allowing their videos to train the next generation of AI models. You can read a recent Tubefilter article here.

Your channel, with approximately 351.3 hours of content, would perfectly fit into our program. Here's what we're offering:

• New revenue stream: Based on your catalog size, we estimate it to be worth around $35,130.24, with a first deal potentially unlocking about $4,215.63.
• No impact on your rights: Your content won't be displayed anywhere, and you'll retain full control over your rights, content ownership, and monetization of your videos.
• Easy setup: The process is quick and seamless, requiring minimal time from you.
Let's schedule a quick 15-minute call to explore this opportunity further; you can find a time https://meetings.hubspot.com/mde-franco/rhei-meeting. Alternatively, you can get started right away by signing up here.

I look forward to connecting with you.

Best,

Marco

Marco De Franco
RHEI Monetization Expert

FIGURE 3.1
Screenshot of an email offer for the YouTube videos to be used for training AI.

Furthermore, whilst the email states that the creators retain the rights over their content, which is true and whilst they own their own likeliness, if the video is realistic enough, then it could potentially conflict with the original videos from which it was generated and therefore encroach on the audience of the original channel. It means that creators could begin to lose their audience, especially if they attract backlash if their audience discovers they are paid for providing their content for training AI. Furthermore, YouTubers such as Kwebbelkop have already created AI-generated videos with a strange AI-generated version of themselves (Kwebbelkop, 2022). The videos are quite frankly broken and uncanny, but whilst this is not an example of theft itself, it certainly paints an interesting view of the future. Furthermore, videos with AI-generated voices that summarise TV shows and films already dominate YouTube and arguably steal views from the original media. However, AI can be used to generate more 'original' content such as original songs with AI-generated videos such as Soviet- and North Korean-themed Penguins marching in time to music, which likely could not have been properly realised and produced without considerable time and money. Yet as creative as these videos can be, they are still made from stolen content that is taken without the consent of the creators.

Finally, when considering AI's influence on film and television, AI can be used to copy dead actors' likenesses by using voice cloning and generated images. This is already being done, notably by a music producer and Doctor Who fan Ian Levine, who spent significant time and resources recreating the lost Doctor Who episodes without permission from the BBC or asking the estates or descendants of the actors concerned (Lungariello, 2025).

The finished episodes often look uncanny, whilst the quality is very poor overall due to lack of concentration on this area.

This contrasts with the CGI Peter Cushing as Grand Moff Tarkin in Rogue One (Gilroy & Edwards, 2016), which was created with the use of both a stand-in actor and footage from Star Wars: A New Hope (Lucas, 1977). A digital mask built from the footage was applied to the actor, and the CGI Tarkin was created with little to no AI involvement. Yet it is unlikely that this will be used much, as Tony Gilroy, Rogue One's scriptwriter, states that much of the process to bring back actors will be taken over by Machine Learning in the future (Lund, 2024). Furthermore, this area formed part of the 2023 Screen actors Guild – American Federation of Television and Radio Artists (SAG-AFTRA) strike, which occurred from July 14 to November 9, 2023, and the 2024 SAG-AFTRA video-game strike. These strikes were intended to provide actors protection against digital recreation and voice cloning as well as other forms of AI used to replace actors. When the 2023 SAG-AFTRA strike concluded, it granted actors protection against AI, requiring studios to ask permission before a digital replica is created of them and, furthermore, paying them for the work their digital replica undertakes (Anguiano, 2023). This also extends to extras and background performers. At the time of writing, the video game actor strike is ongoing, yet it strives for similar protections from AI.

The idea of AI replacement actors has already been tried with a rather uncanny and crude line of AI students edited into a scene of Prom Pact, the American romantic comedy television film (Anarumo, 2023), where extras could have easily been used. Whilst CGI actors for wide shots and fully CGI characters are commonplace within the film industry, these are often at present painstakingly made by artists, whereas it is more likely in the future to be undertaken by Generative AI. If this is done expertly then it is likely that audiences will not even notice, therefore exacerbating the problem of unseen theft. Overall, the problem of the unseen theft of art is unlikely to disappear, and it is up to the artists and their supporters to fight for the protection of their art. Thus, technologies such as Nightshade and Glaze (offensive tools that artists can use as a group to disrupt models that scrape their images without consent) must become commonplace as well as other anti-AI tools. Furthermore, legislation must be published to give people and artworks greater protection against AI. Governments must also hold tech companies to account for their complicity with theft.

Conclusion

Large Language Models are currently treated with too much respect and hold unjustifiably high reputations for being what are essentially glorified guessing machines. The human connection with these models means that traditional testing is no longer as effective. Therefore, tests for sentience and humanity need to evolve as LLMs become more and more advanced, yet this must also come with the advancement of human understanding of these AIs. With this education comes a massive need for a focus on AI learning, the theft involved during the training stage and the requirement for the protection of art.

4

The uses and dangers of LLMs

Introduction

This chapter begins by examining concerns about LLMs, focussing on some of the dangers, promises and issues of social justice. The second section of this chapter examines the impact of these models on humanity and society, arguing that there are ethical and privacy issues that need to be addressed such as data scraping, stolen art, copyright issues and the lack of safeguards in AI image generators. The final section offers an overview of some of the recent cases of AI abuse and suggests ways in which it might be possible to anticipate the future and find ways of developing methods to minimise the effects of such abuse.

The dangers of Large Language Models

One of the challenges of examining the dangers of LLMs is that there tend to be polarised views about their use. There are people who use ChatGPT with little thought to the consequences of using it, and others that are so against it and so concerned that they refuse to engage with any kind of LLM. For example, a friend told me quite blithely about how she needed to contact her local doctor through ChatGPT to provide her details, describe her symptoms and gain an appointment to see them – not realising she was feeding highly personal information into a system for which there are few safety measures, due to ChatGPT potentially making medical errors. This polarity is not only seen in everyday life, but it also is seen in academia. For instance many social scientists have embraced the use of LLMs for sorting documents, annotating texts and summarising research papers, yet even in doing this there are difficulties, which include

- Different LLM models yield different results (Reiss, 2023; Rytting et al., 2023)

DOI: 10.1201/9781003622949-5

- There is little, if any clarity about what safely measures, if any, are in place and what it is the models have been trained on
- It is evident that different results can emerge when the same data set is repeated – either because of changes in the LLM or the updates that have occurred since the first testing (Ollion et al., 2024)
- LLMs appear to perform best in the English language (Lai et al., 2023)
- Due to issues of privacy and intellectual property, only particular types of data can be used in LLMs, as Ollion et al. (2024, p. 4) note.
   OpenAI claims that it does not "use content that you provide to or receive from our API [...] to develop or improve services" (https://openai.com/policies/terms-of-use). But this does not mean that they will not do so in the future or in another way.

It seems until we begin to confront the dangers and abuses of LLM technology, it will not really be possible to harness its positive power effectively.

Dangerous tech

To refer to LLMs as dangerous tech might seem quite stark, but as Vallis (2025) notes, Generative AI is neither artificial nor intelligent. The dangers in LLMs then relate to the undermining of academic integrity, the amplification of bias and the reinforcing of stereotypes.

The use of LLMs to summarise articles and analyse data may be useful, but it also raises questions about whether this is classed as genuine scientific work. Journals and publishers have banned the use of AI-generated images and graphics, and many have restricted the use of AI writing tools because of concerns over counterfeit research findings. Baker and Hawn (2022) note that AI systems amplify bias and reinforce stereotypes, as is discussed below.

A further danger is the reinforcing of stereotypes, which is seen particularly in AI-powered facial recognition software. Sandvig et al. (2016) argue for the need for algorithmic ethics in order that analysts can determine if an algorithm is illegal or unethical. In practice this means understanding the details of computer codes. However, they argue that the term 'algorithm' is problematic as it can refer to everything from computer programs to data as well as the software used on Google. Sandvig et al. suggest that algorithmic ethics requires an analysis of both the composition of the algorithm and its consequences. They argue for three reasons that it is vital to have an ethical stance towards algorithms:

- Algorithms have an ability to delegate authority, resulting in algorithmic design being distanced in space and time.
- The rise of network control because of the speed and rapid distribution of data has shifted the balance so that 'the technology's diffusion changes from that of an inventor hopefully sending

products out into the world to that of puppeteers whose movements remain always connected to their performances' (Sandvig
et al., 2016, p. 4985).

- The permanent stabilisation of algorithms is resulting in ethical
  problems such as problematic and biased facial recognition software.

This work suggests that there is more work to do in terms of detailed ethical
understandings of algorithms, as well as understandings of the social and
broader technological impact of their use. An old pre-algorithmic example
that perhaps sums up some of the difficulties we still face:

> Hewlett-Packard (HP) suffered a serious public relations crisis when it
> was revealed that its implementation of what was probably a bottom-up
> feature-based face localization algorithm (Yang et al., 2002) did not detect
> Black people as having a face (Simon, 2009). Cameras on new HP com-
> puters did not track the faces of Black people in some common lighting
> conditions. In an amusing YouTube video with millions of views, Wanda
> Zamen (who is White) and Desi Cryer (who is Black) demonstrate that
> the HP camera eerily tracks Zamen's face while ignoring Cryer, lead-
> ing Cryer to exclaim jokingly, "Hewlett-Packard computers are racist"
> (Zamen, 2009).
>
> *(Sandvig et al., 2016, p. 4973)*

A more recent incident has been the use of facial recognition technology in
policing, resulting in wrongful accusations in Detroit (Hill, 2022). Wrongful
arrests of African-Americans have been common across the US because
of the use of poor-quality images from grainy surveillance cameras, then
exacerbated by poor algorithms. The reason that this technology is racist is
because it is trained on old categorisations and was based on the construction of what is deemed to be an average human face, which, as Crawford
(2021) notes, is epistemologically dubious and has clear racial bias. As Lu
and Qiu note:

AI-powered FRTs are among the most problematic modes of extracting
data from

> Southern populations due to its invasiveness, automated nature, deep-
> rooted racism, and numerous flaws that are technological as well as ethi-
> cal and legal.
>
> *(Lu & Qiu, 2022, p. 769)*

The dangers, it would seem, are increasing, especially as algorithmic bias
continues apace, with little seeming to be done to address this problem.

Algorithmic bias

In the past algorithms have been assumed to be objective, but it has become
increasingly clear that the biases that are inbuilt are based on the biases of
those who coded them in the first place. Baker and Hawn note:

Examples of algorithmic bias cross contexts, from criminal justice (Angwin et al., 2016), to medicine (O'Reilly-Shah et al., 2020), to computer vision (Klare et al., 2012), to hiring (Garcia, 2016). These limitations appear – and are particularly salient – for high-stakes decisions such as predicting recidivism (Angwin et al., 2016) or administering anaesthesia (O'Reilly-Shah et al., 2020).

(Baker & Hawn, 2002, p. 1053)

Bias is a preconception about a thing, person or group. It means that someone holds a preferential perspective at the expense of (possibly equally valid) alternatives. Bias can cloud judgement and lead people to see what they expect or want to see, which may or may not be what the data suggest. A helpful way to understand bias in relation to algorithms is provided by Suresh and Guttag (2021), these being historical bias, representation bias, measurement bias, aggregation bias, evaluation bias and deployment bias. These six types are explored further in Table 4.1 below:

Baker and Hawn (2022) studied algorithmic bias in education in areas from race/ethnic and gender bias to the less studied areas such as socioeconomic status, military-connected status and disability. What is notable about their findings is the large-scale unknown biases that were evident in studies, along with the areas that have not been examined, such as age as a factor in undergraduate and postgraduate education and professional learning. The authors ask difficult questions such as why Latinx learners or Spanish-speaking learners are treated as monolithic groups and why all learners from China are treated the same (p. 1074). The difficulty, it seems, is that categories are developed from political rather than cultural distinctions. The way forward they suggest is to improve data collection, tools and resources as well as create more openness. For example, openness could be improved by ensuring that journal guidelines require analysis for algorithmic bias as well as broadening the community of people who are working to solve the problems of bias in algorithms; bias is often not noticed by data scientists.

Alleged algorithmic empowerment

What we mean here is that there are technology companies who are promoting their goods with an underlying empowerment rhetoric. In practice alleged algorithmic empowerment is the marketing of goods in ways that prompt the buyer to believe that by buying they are ensuring empowerment and promoting equality, particularly for those with darker skin tones and/or who live in the Global South. An example of this is the study undertaken by Lu and Qiu (2022) who examined Transsion, a Chinese company that has developed facial recognition technology for smartphones. Although their study focussed on an examination of Transsion's patents and patenting practices, they also examined the empowerment rhetoric associated with software designed to recognise darker skin tones. The camera in Transsion's

TABLE 4.1

Forms of bias (based on Suresh & Guttag, 2021)

Forms of bias	Description	Example	Related works
Historical bias	Where data is measured between the worlds as it is or it was, which results in a model that produces harmful outcomes because it was based on data from a particular decade	The use of word embedding in natural language processing which results in biases about women and ethnic minorities	Friedman and Nissenbaum (1996).
Representation bias	When some part of the population is underrepresented in a sample	ImageNet's data set has over 1 million images but 45% of the images were taken in the USA	Shankar et al. (2017)
Measurement bias	This occurs when particular variables are chosen	A model predicting violence or arrest may be biased when violence has been recorded for only Black students	Ensign et al. (2018)
Aggregation bias	Populations are combined in the same model, which means the model does not work for all of the groups	By combining social media posts and training on a nonspecific model, which can result in perceived aggression where there is none	Patton et al. (2020)
Evaluation bias	The data used for the evaluation does not represent the population on which it is being used	The use of facial recognition software on images of dark-skinned woman were found to be inaccurate in detecting gender or smiling direction	Buolamwini and Gebru (2018)
Deployment bias	This is where there is a difference between the way a model was designed to be used and the way it is used in practice	A model designed to predict a prisoner's likelihood of reoffending but are used instead to determine the length of sentence	Collins (2018)

smartphone which recognises darker skin tones has been a key selling point, as well as its rhetoric of empowerment. The empowerment rhetoric centres on Transsion's argument that their software applies beautification algorithms so that skin tones can be beautified properly, the website explains:

> AI-Powered Computational Portrait Engine delivers personalized portraits tailored to regional aesthetics and cultural preferences. It enhances skin tones, textures, and beauty features, ensuring every photo captures your unique beauty perfectly.

> *(Tecno Mobile, 2025)*

Yet whilst this software may begin to decolonise AI, as Lu and Qiu point out, the scale of the data set and the ties with China are of concern in terms of whether they will lead to new forms of surveillance and exploitation. Furthermore, it is not clear how and if the data extracted will be used, whether it will be available for Western markets or indeed if it will be used for anti-Black surveillance purposes.

An additional concern is what is termed 'data colonialisation' by Couldry and Mejias (2019) This is described as the process by which governments, non-governmental organisations and corporations claim ownership of and privatise the data that is produced by their users and citizens. In practice, this means normalising the exploitation of human beings through their data; they argue:

> In the hollowed out social world of data colonialism, data practices invade the space of the self by making tracking a permanent feature of life, expanding and deepening the basis on which human beings can exploit each other.

> *(p. 344)*

The notion of locating the capture of personal data from a colonialist stance is an interesting one, but the authors tend to deal with colonialism lightly. We suggest then that to a large extent this view has been somewhat overtaken by work such as Zuboff's Surveillance Capitalism. Further explorations of the transparency in AI technologies in continents such as Africa, where concerns about data privacy and security are prevalent, are perhaps more helpful (Ade-Ibijola and Okonkwo, 2023). Training algorithms against bias and harmful forms of facial recognition technologies is vital so that it becomes possible to ensure underrepresented groups are included and greater equality and equity are assured.

Justice and integrity

Much of the digital injustice happening in the 21st century is occurring through relational technologies: knowledge and news are spread through tweets, memes, shares and likes. Social media, whether X (Twitter), Snapchat or Tiktok influence our lives, often in ways that are not realised. The result is that social opinion becomes knowledge. Thus, truth(s) are socially engineered by the prolific voices of social media because there is no restriction on the socialisation of digital technologies. The result is widening inequality – digital, social and political. Further layers of this are seen in the global increase in the invention of fake news; when state, military and political leadership prohibit or suppress social media usage, such as in China or North Korea; and in the West suppressing or hiding news coverage of legitimate demonstrations and miscarriages of justice. This widespread injustice also hides the plight of those in poverty, being treated inhumanely in prison, being discriminated against in covert ways or being arrested and tortured for crimes they did not commit.

There are also questions about human labour and the questionable manufacturing conditions in China (Qiu, 2016) and Kenya (Lee, 2018). The hours are long, and the activities are repetitive and routine for low wages. Qiu (2016) compares the 17th-century transatlantic slave trade with the iSlaves of the 21st century. The iSlaves Qui defines are the Chinese workers at Foxconn, an electronics company whose largest buyer is Apple. Qui notes that student interns are paid below the minimum wage, legally supported by the Chinese Government. The challenge from Qui is that there is a need to rethink our relationship with digital technologies and examine in depth the impact of LLMs on humanity and society.

Dupery and fraud

Large Language Models are fast becoming an effective tool for committing a variety of fraud. There are the obvious uses, such as using ChatGPT to generate more convincing scam emails and cold calls. Yet it becomes even more dangerous when it comes to voice cloning. This is commonly used to make parodies of existing songs in funny voices such as having droids from Star Wars singing sea shanties (Mr.Rand0mStuf, 2024), but there is a much more nefarious use. Various banks use voice authentication, and the phrase 'My voice is my password' can be used when accessing your account by phone call. A criminal can make a scam call to get a sample of a voice, then run it through voice cloning software to obtain a voice print before finally using this to access the bank account (Vahl, 2024). Although this can be mitigated by disabling voice authentication, if such voice authentication becomes widespread, then voice cloning still presents a clear danger. Furthermore, the technology exists to autogenerate podcasts and other media which can be exploited for monetary gain or to promote fake or defective products or to share fake news (Note GPT, 2025). Given the already growing influence of podcasts, especially on the younger generations, the idea of AI application to improve the believability of the scams means that they become an even greater problem.

When considering the more common uses of LLMs such as for the generation of text, these can be misused to create fake documents such as bank statements, legal documents and perhaps even court orders, which could be used for blackmail or in the deception of a jury (Thaldar, 2025). Yet another use is when applied to phishing emails and scam call scripts, leading to more convincing scams. When this is coupled with AI image generators, this can lead to significant fake advertising and other fabricated content. One such example is a game trailer released by Snail Games that is little more than AI images and videos mashed together with little resemblance to the actual game. This action caused a major backlash within the gaming community, and the developers were quick to call out their own marketing team for the blunder, stating that they had no knowledge of it. Yet with the improvement of AI technology, there is the ability for generated content to become increasingly convincing.

However, this technology can also be used for good, as demonstrated by the creation of Daisy, the 'AI Granny' developed by O2, who is designed to waste the time of scammers by prattling on and on with random conversations, as well as to gain possible information on the scammers to assist law enforcement. Daisy can also provide fake phone and bank details if required to waste a scammers' time further. A further example is Apate (Apate.AI, 2024) which uses two chatbots exhibiting unsteady old men personas called Ibrahim and Malcom (Shepherd, 2024). There is also an AI email replier system known as Re:scam, developed by Netsafe. If a scam email is forwarded to this bot, it will reply for you, providing the scammer with significant misdirection as well as false information (Netsafe, 2017).

Yet on the horizon is an even bigger concern, which is the increasing capabilities of LLMs to create deepfake technology and AI-generated selfies, the latter of which can be used to fake selfie verification on social media apps and for fake ID cards (AU10TIX, 2024). Yet deepfakes are the bigger concern, as they can be used to make politicians appear to say almost anything that is a massive contributor to fake news. Furthermore, this can be an even bigger concern when applied to court cases as the proliferation of fake evidence increases the potential of innocent people going to prison. One such example is demonstrated within the TV show The Capture (Chaman, 2019), where deepfake technology and voice cloning were used to frame multiple people as well as to sow confusion and doubt, thereby impeding the course of justice.

Impact on humanity and society

The extensive use and integration of LLMs into society and humanity means a global transformation within the digital age. The use of ChatGPT and Llama (another opensource AI model), particularly throughout the education sector, means a significant rise in cheating and academic misconduct due to the usage of LLMs in essays and homework. This has also caused a rise in false accusations of misconduct, due to the poor ability of AI detection systems, culminating in the erosion of trust between students and academic institutions. This also means there will probably be a rise in the number of more face-to-face examinations and vivas or more increased monitoring of student online conduct. An even more concerning factor is ChatGPT and Copilot's abilities to generate code, which enables almost everyone to be able to code which means a lowering in expertise needed for creating programmes. Whilst this can be beneficial for app development, it also means a greater amount of malware creation and deployment that could potentially cause damage to both infrastructure and people. One of the most common uses would be the generation of scam email bots as well as chatbots that lurk on illicit websites that are designed to scam

Furthermore, with the rise of AI-generated content, specifically image generation, there comes the increased usage of AI in advertising and misinformation campaigns as well as fake news. This means a rise in people being taken in by more and more convincing scams such as the AI Elon Musk giving out free bitcoin, as well as news stories, rewriting the world perception around people. Therefore, there is likely to be an increase in the polarisation of views across the world as AI influences social media one way or another. This is particularly problematic with the deepfakes discussed earlier, as political campaign videos could be generated to support or smear political candidates, therefore influencing elections. Currently this technology is being used rather harmlessly in the form of meme Minecraft videos featuring British politicians like Keir Starmer, Nigel Farage, Boris Johnson and Rishi Sunak for entertainment (Thermo, 2024). Yet there is always the potential for these AI systems to be weaponised for use in the political world. Hence, the truth will become increasingly harder to discern.

A more curious case of AI's impact is its use in dating, beyond the traditional catfishing and AI filters. The rise of LLMs has led to the invention of applications such as RizzApp. Rizz is defined as charisma when it applies to flirting. These apps are designed to help everyone to seemingly gain more charisma, yet whilst this may be suitable for online dating and texting, it may become problematic when people lack the same charisma in person. Furthermore, given the AI may lack context for the specific relationship, there is the possibility that it can hinder rather than help, especially if it generates problematic lines of conversation. Furthermore, one prevalent prompt used on Hinge and Bumble is 'Who do you go to for advice?' with one of the more common answers being ChatGPT. This is concerning for a number of reasons, as when talking to AI instead of friends, AI lacks the context and knowledge of the person providing the information, in contrast to their friends who know more of their nature, characters and actions. Furthermore, the AI can be influenced by the user introducing bias into the prompt, leading to them getting misled or becoming a victim of confirmation bias. This overreliance on AI's dictating romance and dating can potentially lead to the automation of a relationship and matching people who are not suited, due to preferring the AI version of the person over the real one.

When considering inequality within AI, there arises the problem of wealth inequality, as suggested by Mahajan (2023), who argued that the job losses caused by AI automation, with little chance of much reemployment, changes the idea of social mobility. Yet the wealth economy will widen not only because of job loss but also due to investors who have shares in AI companies with the likelihood of their earnings increasing vastly. Furthermore, the AI advancement of these companies means a shortage of certain computer components, driving the prices up. This is already happening with Graphics Processing Units (GPU) in computers, with prices increasing considerably in the past few years (Kan, 2021), and the price rises have affected the gaming and computing community, making it more inaccessible. This suggests significant economic change, coupled with the fact that AI integration can produce significant breakthroughs in multiple areas of industry and increase human capability in the workplace through AI-assisted tasks, which indicates that AI is already causing significant economic upheaval.

Additionally, with the growth of AI, there is an increasing demand for energy, leading corporations such as Google to begin investment in energy sources such as nuclear reactors. Whilst this will potentially increase the renewable energy sources in the world, this may cause AI to be prioritised over people and their resources. A further consideration is the bias within AI models since they are often trained on an imbalanced dataset, and attempts to mitigate this problem have been problematic. Moreover, AI systems typically have massive datasets and/or have blackbox systems that do not allow anyone to access how the AI makes decisions, in the interest of so-called user privacy. This means that fixing the bias within datasets is extremely difficult, and therefore, it is likely that this problem will persist.

Finally, when considering AI replacing human actions, this means that common jobs would feel more dehumanised. One example of this is the utilisation of automatic till machines which, though successful, grew unpopular and led to a greater preference for humans at the tills (Dejevsky, 2023). Furthermore, this also applies to AI art, which people refer to as 'unnatural' and 'soulless'. Art comes from a deeply emotional place; LLMs cannot replicate emotions and therefore cannot channel them into art. This lack of emotional response is an even greater problem because LLMs are often unlikely to admit mistakes or say, 'I don't know,' or even provide sources that do not exist. Furthermore, if you provide them with an artistic work, they are unlikely to say, 'I don't like it', as it will not criticise or be negative unless explicitly directed to be. However, this lack of human touch does convince users of its objectivity, which leads users to put more trust in the LLM, as whatever you do, it will not judge in the same way as a human, but in doing so, it is likely to be biased towards their user. For example, you can instruct ChatGPT to justify a crime or even perform as a lawyer, as has happened in a court case with an AI-generated lawyer (Neumeister, 2025). Whilst the AI lawyer was dismissed as not being a valid legal defence, it does show the level of reliance that certain users place on AI-generated tools.

Furthermore, when considering AI in the legal world, paralegals and court Clerks are likely to use ChatGPT to help summarise legal documents, which could lead to further problems if they miss something, meaning that LLMs can have a profound impact on the justice system.

Overall, the impact of LLMs on both humanity and society is an issue that cannot be ignored, as it has the potential to affect whole sections of society, as well as cause and solve a huge variety of humanity's problems. Therefore, it is an issue that should be treated with more caution and respect than it is currently being given. A significant amount of the issues with LLMs are relatively unknown to everyday users, such as their extensive training process, which causes huge energy resource consumption, as well as when the user subsequently uses the trained LLM. There are also a significant number of more nefarious cases of LLM misuse and abuse that are occurring in the more hidden corners of the internet and the dark web.

Other cases of AI abuse

A major case of AI abuse growing in online spaces is its usage in sexual forms, for example, in generating deepfake explicit photos of celebrities which could be used for blackmail. Generating these images was recently criminalised in the UK (Ministry of Justice, 2025). However, it is currently still legal for AI image generators to circulate explicit photos of minors with little to no restrictions, and these often carry only a slight warning concerning their problematic content. This is highly troubling for a number of reasons, not least that the AI may have obtained many troublesome images during its training process due to the lack of preventative measures to stop this occurring. Another growing concern is AI usage in E-dating, where an entire profile can be generated to attract fans of all ages and then used for extortion or blackmail. Furthermore, it is sometimes minors as young as 14 that utilise these accounts to 'e-date' and scam other minors or adults, with the goal purely being for monetary gain with little regard to consequences or safety. There is also concern surrounding sites like Character AI and AI dungeon, which allow the creation of bots at any age, some of which allow the sending of explicit AI images. Additionally, these sites have also led to the manipulating of young people and the taking advantage of those with insecurities or mental issues, leading to horrifying consequences, such as the teenager who committed suicide due to their reaction to an AI version of Daenerys Targaryen from Game of Thrones (Carroll, 2024).

Deepfake technology and the relentless growth of AI capabilities such as its ability to generate ever more realistic videos will lead to significant issues in the future. Whilst most may use these harmlessly, such as producing cats walking or people dancing (Open AI, 2024). There is the concern that such videos could equally be used for fake evidence, blackmail, framing, fraud and other forms of abuse.

Despite our concerns about LLMs, David Burden offers a more positive stance:

EXPERT REFLECTION
David Burden, *CEO, Daden Ltd*

'ON THE WAY, BUT NOT THERE YET... '

The introduction of Large Language Models (LLMs) has led to significant advances in our ability to create useful and powerful chatbot and conversational AI applications. Tasks that 5 or 10 years ago may have taken separate months-long projects to deliver can now be achieved by just asking the bot to do things in a particular way. In my experience to date, LLMs have been most effective when being asked to do something that is highly narrative, rather than trying to get them to do something that is highly prescriptive. If you're prepared (and able) to just go with the flow of the conversation, then a lot can be achieved. Of course, there are problems around hallucinations (all the more reason to keep discussions at a narrative/conceptual level), and the bot does tend to reflect the biases of its training data. I've also found, along with many others, that the bots have a tendency to please, to give the user a positive answer (even if it means lying) and to neatly wrap tasks up with no loose ends hanging – even if that's not what reality is like.

There is a danger that LLMs are seen as being *the* solution to conversational AI, whereas I suspect they will be just one part of the solution. In particular, I believe that knowledge graphs, whilst pre-dating LLMs, are still a very powerful tool for conversational AI, as they offer some of the semantic awareness and explainability that LLMs lack.

Knowledge graphs can help create very human-like conversations by having the bot talk about and around the topic for as long as you let it – just like many real, somewhat one-sided human conversations – and helping to solve the so-called discovery problem. Having LLMs working in concert with knowledge graphs is an area of much active research. Another area in which I have a keen interest is in embodying bots within virtual environments – where they can have a grounded, embodied experience to help develop their own intelligence, and where we can interact with them on a level playing field. As to where this will all go, I often talk about two key models. I am inspired by the dæmons from Philip Pullman's His Dark Materials trilogy where the bot is the fount of all knowledge and provides both 'best practice' and unbiased information and wise guidance and mentoring – and how we produce not just artificial intelligence, but also artificial wisdom is a keen interest of mine.

The other is inspired by the portraits of previous headmasters and headmistresses of Hogwarts in Dumbledore's study. He is able to talk to each character in the portraits, a one-to-one conversation with a previous post-holder, with a digital immortal if you will, but each conversation reflects not only the expertise and experiences of the person but also their biases, knowledge gaps, enthusiasms and personal foibles – it is a warts-and-all experience. When building bots, we need to be clear about which one we are building, and the user needs to know which they are using.

Conclusion

There is no doubt that LLMs are able to enhance and improve human actions, but there is a significant amount of ignorance among the users of LLMs who seek to use the technology because it is in vogue or being used for fun or to save money, without thinking about the consequences. The lack of limits and restrictions means that the potential for LLM misuse is enormous, and therefore, there needs to be an increase in user responsibility as well as more stringent laws and restrictions implemented. Overall, whilst LLMs are a fascinating tool, users must remember that they are only a tool and should be treated as such. People need to be wary of how they use LLMs and the effects on the environment, resource consumption and other users. Whilst LLMs are capable of tremendous achievements, they have been built predominantly from stolen data, require an extensive amount of energy and water consumption and are far from infallible.

5

Climate change

DOI: 10.1201/9781003622949-6

Introduction

This chapter will explore the environmental, social and psychological impact of AI. It begins by offering an overview of the concerns relating to AI and climate change, in particular examining the effects of computer systems on the environment. It then examines power consumption, water consumption, CO_2 emissions to AI and then covert surveillance. The final section will examine the drive towards nuclear options and then analyse the positive and valuable uses of AI which may also minimise environmental effects. It explores the market impact of AI, makes recommendations about what corporate and political responses there should be and finally suggests models and mechanisms that should be put in place.

Understanding AI and climate change

There are a range of views about what should be done about caring for the environment, with some people arguing change must happen but that it is too expensive, whilst others suggest that adaptation rather than mitigation is the only realistic way forward. A somewhat cynical stance is that the planet will be burnt up by 2065 so that whatever we do now will have little, if any, impact. Those who take this view encourage the people of 2025 not to have children but also seek to preserve the environment as much as possible in the hope that their cynicism is misplaced. Mann has argued that we are facing a new climate war because of the number of sceptical stances, he explains:

> The most hard-core contingent - the deniers - are, as we have seen, in the process of going extinct (though there is still a remnant population of them). They are being replaced by other breeds of deceivers and dissemblers, namely, downplayers, deflectors, dividers, delayers, and doomers-willing participants in a multipronged strategy seeking to deflect blame, divide the public, delay action by promoting "alternative" solutions that don't

TABLE 5.1

The effects of computer systems on the environment

Type of effect	Definition	Example	Effect	Related work
Computational	The development of computer hardware.	Graphics Processing Units.	Manufacturing using energy and water.	Kuo et al. (2022)
Application	AI systems affecting the environment by the applications they enable.	Increase in productivity of carbon-intensive industries.	Increasing use of oil and gas.	Greenpeace (2020)
Systemic	AI efficiency results in rebounds costs and production of CO_2.	Self-driving cars.	Reduction in public transport.	Kaack et al. (2022)

actually solve the problem, or insist we simply accept our fate-it's too late to do anything about it anyway, so we might as well keep the oil flowing. The climate wars have thus not ended. They have simply evolved into a new climate war.

(Mann, 2021, p. 45)

Mann's perspective, some years ago, remains relevant, especially as the 2025 US Republican government has changed the government stance on greenhouse gases, ignoring any argument for the need for mitigation. A further example of the difficulties can be seen in The Global Warming Policy Foundation in the UK, founded by former Conservative Chancellor Nigel Lawson. The group promotes a sceptical stance towards climate change, and this is fostered through a strong group within the Conservative party in the UK.

Some of the questions that need to be considered are not just how green businesses, universities, colleges and other institutions are, but how ethical their practices are, even those that may be hidden, such as the source of goods and hardware. For example, there is little exploration of the ethics of computer hardware manufacture, and much of the hardware is bought because it is cheap rather than ethically sourced. Kaack et al. (2022) developed a framework to distinguish between the computational, application and systemic effects of computer systems, which we have summarised in Table 5.1.

The materials used for buildings in terms of the extent to which they are locally sourced and environmentally friendly need to be considered, as does the heating and air conditioning used within buildings and generally the extent of our use of renewable energy. To date relatively few institutions and businesses use solar power, ground source heat pumps or wind power. Other areas that need to be addressed include purchasing fair-trade food, sustainable fish and locally sourced food and increasing our use of fair-trade technology. Wesley Goatley offers his perspective on some of these issues.

EXPERT REFLECTION

Wesley Goatley, *Programme Director of Interaction Design &*
Visual Communication, University of the Arts London.

ENVIRONMENTAL, SOCIAL AND PSYCHOLOGICAL IMPACTS OF AI

Having meaningful conversations around climate responsibility in using AI systems is made considerably more complex due to the opacity of these tools and how they're deployed. For example, asking a question such as "when could the energy cost of generating a 500-word text using a Large Language Model be less than the energy consumed by writing 500 words on a computer?" has an almost infinite number of variables: does the output from the generative process need editing, and if so, how much, and what resources would that take? Did the prompt deliver the intended results the first time, or were multiple prompts used? Where is the *likely* location of the data centre architecture where the computation took place, and what's the *estimated* cost of energy at that time and place?

Away from the infrastructural dimensions, the individual circumstances of the writer are important: is the person who'd write the text instead of generating it doing so in their first language? Are they a fast typer? Do they have any impairment that might slow their reading and writing? Each of these questions can wildly impact the overall calculations of energy use for this individual task, and so attempting to make generalised assessments on AI usage with such granular variables is always going to be folly.

Though a general assessment is impossible, grappling with this question itself is valuable and offers opportunities to centre personal agency and foster socio-technical knowledge. When considering the energy cost of using these systems, one productive approach is to reduce the number of variables at hand by opting to run AI tools locally on a home machine and then consider the impacts from there. Locally hosting an AI model generally means that the only data centre interaction required is for initially downloading the model (whose training has, of course, consumed an incredible amount of resources), and all further operations are offline from that point onwards. The privacy benefits of this aside, it also gives the user an opportunity to do their own, less speculative investigations into energy use. Free open-source software can quickly and accurately report how much power is being used to run the local-hosted model and its outputs, versus writing the same amount by hand on the same machine, and local estimations of carbon cost per kWh can typically be made with a high(er) degree of confidence. From there, the user can make a series of active decisions, such as changing to a more efficient model for the specs of their computer or changing the type and style of model to better suit the type of queries that they tend to make. Setting aside questions of embodied carbon in either the model, the data centre, the laptop, etc., this process becomes one where the user gains real knowledge of the tools, their capacities and the nuances of the carbon cost of their use. Such approaches are a step towards making more informed ethical and critical judgements as to when to use, or not use, these tools.

Authors such as Mateos-Garcia et al. (2024) suggest that there could be some positive applications for AI, which include:

- The reduction of greenhouses gases through innovations such as smart cities and smart buildings
- Better weather forecasting may improve the ability to gain sustainable energy
- The creation of new energy sources such as nuclear fusion
- Scheduling of energy use so that is more effective and sustainable
- Changing transport to meet demand and changing freight routing

It is also worth noting that studies are being undertaken worldwide to model the impact of AI on climate change in order to show what is occurring in an attempt to mitigate it.

Power consumption and the climate impact of AI

The main difficulty with the new AI systems such as ChatGPT and DeepSeek is that the training of these models is highly resource intensive, particularly in relation to water and energy consumption. A powerful example is that the creation of one AI image takes up as much power as it does to charge a smartphone (Luccioni et al., 2024). This section examines the research to date that has sought to understand the impact of AI on the planet.

Energy consumption

Most people have several computers in their homes, and whilst these do have an energy cost, the cost of training Neural Networks for Machine Learning consumes huge amounts of energy in comparison.

Energy consumption also occurs when the raw material such as nickel and cobalt is extracted to create machines in the first place. Furthermore, a recent study has suggested that global AI could consume 85–134 TWh of electricity in 2027 (De Vries, 2023). Alphabet is an American multinational technology company and is the world's third-largest technology company in terms of revenue. In February 2023 Alphabet's chair indicated that interacting with an LLM could probably 'cost 10 times more than a standard keyword search' (Dastin & Nellis, 2023) – since a standard Google search is alleged to use 0.3 Wh of electricity, which is equivalent to roughly 0.2g of carbon dioxide.

Water consumption

Water consumption can be referred to as *water withdrawal*, where fresh water is taken from surface water or the ground, or *water removal* which is where water is removed from the water environment through evaporation, crop consumption or incorporation into products and therefore is no longer available for reuse. The way in which water is used in AI is for data centre cooling, server manufacture and generating electricity. For example, high water consumption is required in order to cool its data centres. Li et al. (2023) state

> ... training the GPT-3 language model in Microsoft's state-of-the-art U.S. data centers can directly evaporate 700,000 liters of clean freshwater, but such information has been kept a secret. More critically, the global AI demand is projected to account for 4.2–6.6 billion cubic meters of water withdrawal in 2027, which is more than the total annual water withdrawal of 4–6 Denmark or half of the United Kingdom.

> *(Li et al., 2023, p. 1)*

Li et al. suggest the following recommendations:

1) tracking and reporting AI water consumption so there is more transparency about exactly what is being used
2) scheduling AI training to reduce the water footprint in terms of time and place
3) avoiding high-temperature hours of the day to ensure water usage is effective

There is a need to recognise and implement such recommendations in order to reduce AI's water footprint. It is worth noting the sheer amount of water used, as noted below, and that few people worldwide realise this cost:

> ... training GPT-3 in Microsoft's U.S. data centers can consume a total of 5.4 million liters of water, including **700,000 liters** of scope-1 on-site water consumption.

Additionally, GPT-3 needs to 'drink' (i.e., consume) a **500 ml bottle of water** for roughly 10–50 medium-length responses, depending on when and where it is deployed.

(Li et al., 2023, p. 2, bold in original)

CO$_2$ emissions

It is noticeable that there are few studies on the impact of carbon emissions and AI but as Zhong et al. (2024, p. 2977) point out,

> Existing studies are based on micro or macro samples of developed countries, yet there is a non-linear relationship between economic development and carbon emissions, moderated by renewable energy use, urbanization, income.

CO$_2$ emissions take many forms, but in the case of AI, perhaps the most powerful (and often cited example) is that of Strubell et al. (2020) who note that training a single natural processing model can result in emissions of 300,000kg of CO$_2$ equivalent, which, to put it in context, is five times the amount that would be produced by an average petrol-fuelled car over its lifetime.

Whilst AI is seen by some as a means of improving industrial structures and decreasing energy consumption (Chen et al., 2022), it is clear that AI is causing problems. Despite the Paris Agreement of 2015, when 197 countries stated their intention to keep the global average temperature below 2° C, global greenhouse gases continue to rise. Zhong et al. (2024) undertook a study to examine the relationship between AI and carbon emissions. The findings indicate that the impact of AI on carbon emissions varies across countries due to the industrial environment of the country and demographic structures – with lower carbon emissions in places with older populations. Furthermore, the carbon reduction effects of AI only exist in high-income and high-carbon emission countries – which would seem to introduce questions about whether the use of AI for reducing carbon emission in such countries can really be effective in mitigating the high AI use in the first place.

Covert surveillance

Whilst covert surveillance might seem an unusual concern to raise in relation to AI and climate change, de Godoy et al. (2021) note that the hyper-networked society has resulted in a decrease in individual moral agency. An example of this is the implementation of smart meter appliances which do not consider people's views or offer them the ability to decide on the privacy of their data. Le Ray and Pinson (2020) argued that as there is little ethical guidance on the handling of smart meters, this has resulted in low trust, and they suggest that the General Data Protection Regulation

(European Union, 2016) should be used to ensure people's homes are not affected by external interferences. For many people, the use of smart meters is assumed to be a positive step to help to reduce their environmental impact, but few people consider the intrusive effects such meters may have in terms of the secret storing of their data and the covert surveillance of their energy use practices.

Positive uses of AI in relation to climate change

There is currently a growing body of literature that proclaims the value of AI for managing climate change, but many of the promises seem yet to be realised. In 2019 Donhauser suggested that there was a need for ethics specifically applied to environmental technologies before the development of them had shifted beyond our ability to make informed decisions (Donhauser, 2019, p. 192), but we now seem to be in a position where the horse has already bolted. However, attempts are being made to use AI to mitigate the climate effects of AI, which, to some extent, does seem to have rather an irony about it. Some of the suggestions provided by Donhauser (2024) for mitigation include:

1) AI monitoring by using AI for conservation and sustainability such as mapping deforestation, pollution and climate cycles.
2) The use of environmental robots, which are either ecobots or robots made with sustainable materials. One such ecobot is the COTSbot. This bot is an autonomous underwater vehicle designed to eliminate the crown of thorns starfish by injecting it with a toxin. Another example is a tree-climbing robot designed to manage pest control and provide tree maintenance.
3) Using simulation and prediction models of AI, for example for predicting natural disasters and extreme weather by using Machine Learning to analyse large amounts of weather and climate-related data.

Whilst Donhauser (2024) has raised 10 ethical questions about the relationship between AI and the environment, he does not really grapple with issues such as water consumption and CO_2 emissions but rather discusses issues such as the safe use of autonomous functions, the collection of sensitive data and ways of evaluating the harms of AI in relation to stakeholder groups. We argue that what is really needed is an in-depth comparison of the use of AI to support environmental sustainability and the climate impact of using AI for this very purpose.

The drive towards nuclear and new energies

The rise of artificial intelligence means increased energy consumption due to the fact that the more data the AI gains, the more processing power is needed to run the AI. An example of this is that in 2022, 50 prompts took the same amount of energy as running one prompt in 2025 (a prompt in AI is the input provided to a model to elicit a specific response). Furthermore, that is just the energy of running the AI. The energy used to train would be substantially more. ChatGPT-3 took 1287 megawatt hours (MWh) to train, which is the equivalent of the annual energy consumption of 120 average American homes. However, ChatGPT-4 took 1750 MWh or the annual consumption of 160 average American homes (Aibin, 2024). Furthermore, given that AIs are often trained multiple times over and over to ensure optimal results, this contributes to an enormous amount of energy consumption. Due to the current energy crisis, this is causing corporations to turn towards alternative energy sources to support their AI systems. Therefore, Google and Meta are funding nuclear reactor construction with the agreement that they get a certain percentage of the output once completed (da Silva, 2024), whilst Amazon has acquired a nuclear-powered data centre (Sherman, 2024) (Euronews, 2024). Whilst this drive towards a cleaner source of energy is good overall for the environment, if these reactors are constantly used for artificial intelligence, then it raises the concern of prioritising the AI and profit over people, with more energy going to a data centre rather than people's homes, which can become a particularly prevalent problem in times of disaster when energy is scarce.

We argue that tech corporations should have a greater push towards alternative energy sources and towards carbon neutrality. Google has already partnered with Fervo to launch a geothermal energy project to power their data centres in Nevada (Terrell, 2023). However, not many other Big Tech corporations are doing the same. Google are also seeking to use turbines and solar panels to power their data centres. Yet there are limits to this, as data centres do draw a significant amount of energy, and the solar panels and wind turbines are unable to keep up due to the uncertainty of sunlight and wind. Therefore, it makes more sense for Big Tech companies to invest in better battery and energy storage technologies to ensure that the data centres obtain sufficient power. Whilst Tesla has already begun building so-called battery parks for increased energy storage, there have been some significant safety concerns due to them catching fire (Australian Associated Press, 2023). Furthermore, battery technology does require cobalt and lithium, which are extremely polluting to mine. Therefore, the common water reservoir as a battery is a more suitable solution. At times of excess power, water is pumped up to a reservoir on a hill; then, when power is required, it is let down through a hydroelectric turbine to generate power.

There is an unexpected benefit to the increased heat produced by data centres since it can be used to heat homes and buildings. Google has already utilised this ability to heat their own offices and have announced their intention to extend this to cities, which could save on significant heating costs (Calma, 2024). Furthermore, although nuclear reactors have an exceedingly high safety standard due to the previous disasters of Chernobyl, Fukushima and Three-mile Island, the risk exists that the increased demand for energy can lead to possible increased strain on nuclear reactors, especially if they are under the sole control of corporations whose primary focus is profits. This will mean a slight increase in the potential for a nuclear disaster. The question must then be asked, how many slight increases of risk before disaster? Thus, there must be a significant expansion of the agencies that regulate nuclear systems such as the US Department of Energy. These government departments responsible for energy use should ensure that both regulations enforce proper procedures and ensure that energy is used responsibly so that AI generation is not placed above human lives. This should extend to all powerplants where a proportion of their energy is used for artificial intelligence.

Expenses and market impact

The recent rise of cryptocurrency and NFTs over the past decade has led to global price spikes in Graphics Processing Units (GPU) or more commonly referred to as graphics cards. These cards are optimal for performing the simple, quick calculations needed for cryptocurrency, yet these calculations, due to the sheer number, take a significant amount of time. This explosion in cryptocurrency mining has led to a rise of both scalping – a short-term trading strategy where traders aim to make numerous small profits by capitalising on minor price movements within a short timeframe and price gouging, which is the act of increasing the price of a good, service or commodity to a higher price than reasonable, often due to increased demand. Scalping and price gouging occur when traders use bots to buy the GPUs and then sell them on at a higher price to crypto miners and gamers for profit. Nvidia has restricted graphics card sales per customer to try to curb this. Furthermore, third-party sellers of graphic cards have increased the prices, whilst other corporations have elected to buy directly from Nvidia via contract and then set up various warehouses to solely and constantly mine bitcoin over and over for massive profits. This has massively affected the gaming industry and community, as it has driven up the cost of personal computers, especially those used for gaming and game development. This has had a profound impact on small businesses, streamers, gamers and individuals who are forced to spend more on GPUs or to buy a cheaper second-hand one which may have been worn out by crypto mining, meaning a probable loss in quality.

Artificial intelligence works in a similar way to cryptocurrency, using many quick simple calculations that together take hours to train AI, and these are best carried out with graphics cards. This could potentially lead to corporations attempting to outbid each other for the technology and to a surge in mining efforts to keep up with demand. Due to improper practices in rare earth metal mining (Nayar, 2021), this is likely to cause an increased slave labour force and human rights violations. Furthermore, it is likely to also cause similar price surges to the bitcoin mining situation and drive up the prices of GPUs still further, making the market ever more challenging for the average consumer. However, with the invention of Neural Processing Units (NPUs) and Tensor Processing Units (TPU), this is likely to decrease the GPU price somewhat, as large corporations began to change over to these components. This will mean that smaller corporations become more likely to adopt GPUs as a cost-saving measure due to not being able to afford TPUs or NPUs. Therefore, price gouging and scalping of GPUs are likely to be increasingly more common, leading to increasing costs for the individual consumer.

Furthermore, as the gaming industry uses AI to improve graphics and gameplay with advanced features such as raytracing, it will only increase the market price of GPUs and cause higher demand. There is also the idea of AI games such as AI text adventures with dynamic endings made by a customised AI, as well as games with programmable characters such as AI People or Character AI. Whilst these games are not that resource intensive, in the future there is the potential for more advanced AI to be implemented in more AAA games. This will also lead to a significantly increased demand for raw materials for GPU manufacturers, worsening the climate crisis due to increased mining, which again will lead to increased slave labour and human rights violations.

The corporate response

Regarding the price problems caused by scalpers, Nvidia has limited graphics card purchases to one per customer, yet this has had minor mitigation effect on the scalping, leading to Nvidia to increase production of their cards to keep up with demand. Nvidia has also begun producing cards specifically for AI training and learning such as the A100 Tensor Core and the RTX series. These cards can also support running LLMs and other AI systems on personal computers, allowing programmers and AI enthusiasts to run AI systems from their homes. Furthermore, advanced micro devices (AMD's) have also increased their GPU production and developed their own AI cards such as the PRO R9700 to allow AI training and model development. Whilst these cards further enhance the abilities of AI as well as increase the efficiency of training, they do come at a cost, both in price and in precious resources. Furthermore, this will serve to significantly increase demand in these cards, meaning that these corporations can essentially charge what they want for this technology, as the AI companies buying them will pay the price to stay ahead of their competitors and further their AI's advancement.

When turning to the environmental impact, the increased production of these graphic cards is likely to cause further environmental damage, especially given that mining for these rare earth metals is one of the most polluting actions. Furthermore, there is likely to be a significant increase in CO_2 emissions, not just from the increased mining but from the construction of new factories to produce these new graphics cards and AI technologies.

Therefore, there must be increased regulation and an increased push towards net zero to reduce the environmental damage. When considering the current pledges of Nvidia and AMD, there is good progress. Seventy-six percent of Nvidia is currently powered by renewable energy and is on track to be 100% renewable by 2025 (NVIDIA, 2024), whereas AMD's data is less up-to-date yet was reported to be 40% renewable in 2023 (AMD, 2024). However, they are also making efforts to ensure that their company partners are moving towards being renewable, so whilst it seems that AMD and Nvidia are poised to massively increase their carbon footprint through their increased production, there is a significant effort to curb their emissions and reach net zero.

Yet one major limitation of this is the mining of the rare earth metals. These rare earth metals are not rare *per se but* are so named because they are highly difficult to extract and require pumping toxic chemicals into the ground. If this is not done properly, then the toxic chemicals can leak into the groundwater and potentially poison water supplies. Furthermore, these chemicals cause significant damage to the workers, causing significant respiratory, nerve and cardiovascular damage (Nayar, 2021). Whilst tech corporations like Nvidia and AMD are not fully responsible for this, as customers, they have some influence over the producers of these rare earth metals and should provide some encouragement for the mining industry to adhere to correct safety standards.

The government response

According to the UN, 190 countries have adopted recommendations on the ethical use of AI, and the USA and EU have also introduced legislation to curb the environmental damage of AI. However, the perspective of the UN is that these agreements do not go far enough to limit the potential damage that AI could do to the environment. The UN also proposed the following five mitigation measures that countries should adopt (UN Environment Programme, 2024):

- Standardise procedures for measuring the environmental impact of AI
- Regulate AI companies so that they must disclose the impact of their AI

- Increase AI efficiency so that less energy is used
- Use renewable energy to power data centres
- Implement policies on AI to meet environmental regulations

Specific countries have implemented various policies:

United States

The United States has introduced no legislation governing AI relating to the environment at the Federal level. However, at the state level, some states have attempted to regulate this. Only Virginia has had some success with House Bill 2035, which required data centres to report the quarterly water and energy use to the Department of Environmental Quality and make the information available to the public (Gooding, 2025).

However, given the current tenuous state of the United States since the 2024 general election and the extremely volatile actions of the government, both in peddling conspiracy theories (The Associated Press, 2025) and their dubious economic policies, it is likely that the Federal approach to AI regulation will be poor and misguided, if it materialises. However, the Federal did try and ban state-level AI regulation for ten years in a recent budgeting bill but faced major pushback from State lawmakers (Sircar, 2025). Furthermore, the vague and broad wording of this law means that it would also forbid States from regulating social media algorithms and deepfakes. Therefore, it seems that it is more of a deregulation bill than anything else. This comes as Congress attracts significant criticism for failing to pass any meaningful AI regulation. So, as the US Government's position seems to be to deregulate AI, as well as pulling out of the Paris Climate Accord (Daly & Borenstein, 2025), their plan for a Federal-level Green AI policy seems to be no plan. Therefore, the only regulation of AI to reduce climate change will be at the state level, and therefore states need to be able to make sufficient laws in order to implement sound Green AI policies.

European Union

The European Union approved the European Green Deal in 2020, which states that AI should be used for further advances towards sustainability and to reach its own sustainability goals. Furthermore, agreements between various technology companies and the EU culminated in the Green Deal Data Space (Gailhofer et al., 2021) to combine and share environmental data and climate science, as well as allow AI to train on this data so that AI systems could then make recommendations on policies for increased sustainability. However, despite this, research conducted into the EU's policies recommends that the current EU policies need to take a bolder stance on the environmental cost of AI, that there is a lack of environmental considerations, and that a fuller assessment of environmental risks is needed. A case study conducted in 2022 advocated the restriction of AI systems that fail to implement global emissions monitoring or CO2 reduction strategies. A further suggestion was mandated efficiency improvements, as in ensuring that AI algorithm efficiency was improved to reduce energy consumption.

So, whilst the EU has started to make some strides in the process towards reducing AI's carbon footprint, there is still much to be done, and the EU must begin to take stronger stances lest the big tech companies take advantage by leaping ahead before the EU's regulations affect them.

Germany

Germany has had a national AI strategy since 2020, which has included extensive dedicated sections on the environmental impact of AI. These plans have been focused on using AI for energy efficiency, conservation and improvements in education, resource management and economy (German Federal Government, 2020), as well as the need for a reduction in energy consumption to create an environment where AI contributes no carbon footprint. Furthermore, Germany operates the 'AI Lighthouse' programme. This programme provides grants to businesses, researchers and non-profit organisations to utilise AI for environmental solutions and has contributed 70 million euros in funding to this endeavour (Bundesumweltministeriums, 2024). Overall, Germany has made a sound start to their Green AI Policy, and it is hoped that they sustain this trend and improve and update their policies when necessary.

China

Although China is known for being one of the most polluting countries in the world, it has slowly begun a transition towards green energy (Bloomberg News, 2024). Yet, like most countries, this has been slowed by the AI boom. Whilst China is behind in terms of AI development, they have become a global leader in adopting Generative AI and are gaining a growing reliance on the technology. This does come with the cost of increased energy usage, and therefore, China has launched the 'Eastern Data and Western Computing' initiative (Zhang et al., 2024). This initiative aims to transfer China's current data centres to the western regions of China. This is due to these regions being more suitable for renewable energy since they experience more wind and sunlight. Therefore, this strategy aims to reduce the carbon footprint of China's data centres and begin a greater push towards renewable energy that can support AI. Whilst China has made a good start towards a green AI approach as well as extensive ethical AI regulations (Global Institute for National Capacity, 2024), it potentially needs to adopt a more progressive stance on its journey towards carbon neutrality and alternative energy usage.

Conclusion

AI has had a significant negative effect on climate change, and although it can significantly help with the push towards green energy and carbon neutrality, there must be significant regulation, especially around water and energy consumption. Furthermore, the practices around obtaining the rare raw materials for AI technology are extremely polluting, so despite AMD and Nvidia making significant efforts towards CO_2 neutrality, they will also continue to make significant greenhouse gas emissions. When considering data centres, good progress has been made by Google and Meta towards nuclear technology, whilst Google has made efforts to utilise the excess heat and further renewable technology. Whilst the corporate response is slow yet making progress, the government response is highly mixed, with Germany and the EU making good progress towards regulating AI to reduce climate change, whilst China has made some steps and the United States is lagging badly behind. Overall, the start for AI being regulated to reduce climate change is slow and positive but may be close to falling behind; therefore, it requires a careful watch by constituents and academics.

6

The morals in the machine

Introduction

This chapter explores the idea of morality when regarding AI models and their development. It will provide an overview of how immorality can creep into AI development and use, as well as how this problem can easily proliferate throughout the industry. It begins by examining what counts as morality in the world of AI and then provides an in-depth analysis of morality in gaming. The second section of this chapter explores Just War theory, suggesting that the use of AI drones and robotic soldiers needs careful consideration. The final section argues that government stances towards AI are underdeveloped and more work is required in this area.

What is morality in the world of AI?

The idea of morality within the world of AI has always been troubling to researchers, developers and data scientists throughout the technology's development. Traditional AI models have been used for years, with the main moral concerns being the dataset and the use of the AI system. The dataset conundrum has predominantly been the origin of the dataset and the bias within it. A biased dataset leads to the AI results being biased, leading to discrimination and/or neglect of the groups with less data. The solution for this is generally to try and balance the dataset so no group has a sway over the learning of the AI. Furthermore, this ties into the origin of the dataset, where has it been collected from? If the data originates from an application whose user base is primarily European, then it may not make sense to balance the dataset, as it will then be overcorrecting based on the limited data on other ethnicities and will sacrifice the accuracy of the European group. Yet the main concern around dataset origin is that the data has been gathered consensually from the participants with their full knowledge of what it will be used for, as well as the user privacy concerns. Yet the privacy concerns

DOI: 10.1201/9781003622949-7

are often bypassed by encrypting said data, leading to issues of humans not being able to inspect it, as that would breach privacy laws. This then leads to the developers and data scientists being unable to explain the results. This became particularly prevalent when social media algorithms recommend problematic content, leading to the discussions in the US Congress on January 31, 2024, in which the CEOs of Meta, Snap Inc. and TikTok were unable to explain the recommendations of problematic content to children and people at risk.

The recent rise in Deep Learning technology has led to an explosion in AI technology such as Large Language Models and other forms of Generative AI. The most well-known is ChatGPT developed by OpenAI. Yet the dataset used to develop this system was stolen from a wide range of writers and artists. Despite this theft, OpenAI has faced little in the way of repercussions. Furthermore, the extensive usage of these systems creates mass competition, which then led to the creation of DeepSeek and various other LLMs. The issue here becomes that it normalises datasets which violate copyright. Yet a realm which is no stranger to the issues of copyright is the realm of gaming, where player's creations are frequently stolen by other players and used as part of other creations.

Gaming

When considering the realm of gaming, some of the traditional ideas of morality are not often compatible. This is due to the world being often gamified, leading to more simplified spaces. For example, usage of flamethrowers and similar incendiary weapons is banned under the Geneva Convention, whilst in most first-person-shooters, a flamethrower is an available weapon. However, the usage of the Red Cross symbol within video games is banned, which has led to various developers having to remove it from their games, no matter if it is a medical centre in Stardew Valley or on a medical kit in Call of Duty. Furthermore, the morality of actions between games is vastly different, depending on the community. For example, the doctrine of warfare in Space Engineers, a space voxel-based sandbox, is often rather polite such as requiring war declarations whilst in ARK: Survival Evolved, a sandbox game involving dinosaurs, the warfare is a lot more unforgiving, with tribes warring on each other for little to no reason or declaration.

The usage of AI in the world of gaming has mainly been limited to the basic programming of Non-Player Characters (NPCs) as well as anti-cheat systems such as Vanguard, which detects hardware and software that is used for cheating within games. However, some games allow the players to programme their own drones or machines to automate specific tasks. Yet often this programming is so rudimentary that it is not considered true AI.

Yet with the proliferation of Generative AI technology and LLMs, game developers have been using it for basic ideas in early development and for reference ideas before hiring artists to carry out the true work. A notable exception to this is Activision who have used it extensively within the Call of Duty franchise, sparking controversy and significant criticism from their fans. However, a more under-the-radar game utilising an LLM is AI People developed by GoodAI Games. AI People allows players to create a full level environment before populating it with LLM-powered NPCs which can be programmed by the players. These NPCs can also interact with the environment the player has created as well as interact with the player-controlled character. The idea of programmable characters that walk around and interact with the player can be potentially problematic since similar AI character interaction software such as AI dungeon and Character AI have led to suicides (Carroll, 2024) and people getting attached to the AI in toxic ways. The ability to interact increasingly effectively with the AI characters leads to the possibility that this problem could be significantly exacerbated and even more significant human attachment to AI, leading them to prefer them to human companionship. This is already beginning to happen with a select few already (Al-Sibai, 2024).

Finally, when turning to the more simplistic games, such as Chess and Go, a company called DeepMind has created AI systems that can surpass humans. Furthermore, DeepMind's AI excelled even more, with the latest being EffiicentZero, an entirely self-learning AI that learns the rules of the game by playing itself in a simulator and then performs better and better. It has successfully mastered Go, Chess, Shogi and fifty-seven Atari games. Whilst this system is not designed to replace humans, it has led to some players such as Lee Sedol, who played against a predecessor of EffiicentZero known as AlphaGo, to retire due to the blistering skill of the AI (Yoo, 2019) as he felt, 'I could no longer enjoy the game'. This does paint a dark picture for both board games and the future of AI players of games, possibly leading to the decline of either board games or AI players.

If this applied to the more conventional world of gaming, then there would be potentially the same effect. However, there may be a small explosion of popularity within the Esports community as pro-players train against AI opponents. This has already begun to happen with the AI known as OpenAI Five developed by OpenAI beating the professional Dota 2 Esports Team OG (OpenAI, 2019). After this event, OpenAI launched an Arena mode to allow any Dota players who signed up to play and train against OpenAI Five. Furthermore, OpenAI also announced that the AI had a rudimentary ability to collaborate with humans, which does provide an interesting idea in terms of game balance. Traditionally in games, when a player leaves a match in progress, either another player is rotated in from the game queue or the team at a disadvantage gets a boost to their in-game stats. Yet with the development of AI agents such as OpenAI Five, it's possible that an AI player can be added to the team instead. Whilst the idea of adding AI teammates is far from new, they are usually rudimentary and fail to make much difference to the outcome of the game. Yet if AI players can be developed and scaled to be as good as the average player on the team it joins, then it can provide a significant balance improvement to the match.

It is only a matter of time before the AIs can excel at most games, yet this can become even more terrifying when applied to horror games. An early example is Alien Isolation's Alien AI system which appears to learn from player actions. In actuality, Alien Isolation implements two simple AI systems. The first being the director, a system that manages the player's menace gauge, an unseen metric that measures how close the alien is, where the player has line of sight on the alien and finally whether the alien could reach the player quickly. This was designed to keep the alien close to keep the player scared but also to back off when the menace gauge gets too high. This is done to avoid the player getting too scared and allow them to achieve some progress. The other system is the Alien itself. As the player progresses through the game, various additional decision trees in the AI's code become unlocked, allowing it more options and traits, as if it's learning from the player (Thompson, 2017). Whilst all the actions within Alien are preprogrammed, the fact that it feels like the Alien learns is very chilling to the player.

Just War theory

Just War theory is the idea that a war can be fought for right and noble reasons as well as being conducted in a certain way. A Just War is intended to be a necessary evil and must restore peace and justice at its conclusion. The idea of Just War dates back as far as Ancient Egypt (source), yet it was St Thomas Aquinas, a Christian monk and theologian who popularised and set out the main conditions, these being as follows:

- The war must have a just cause and not to acquire wealth or power
- The war must be declared and controlled by a proper authority
- The war must be fought to promote good with the aim of restoring peace and justice when the war is concluded.

The above conditions are known as jus ad bellum (Johnson, 2018a, 2018b) while other Chistian scholars such as Saint Augustine have added further conditions known as jus in bello (Johnson, 2024) which are as follows.

- The war must be a last resort when all peaceful solutions have been tried and failed
- The war should be fought with proportionality, just enough forces to win, and only against legitimate targets

- The good that is achieved by the war must be greater than the evil that led to the war.

The idea of AI used for military purposes has long been considered since the 80s, with films such as *War Games* (Badham, 1983) and *The Terminator* (Cameron, 1984). AI has long been utilised in both policing and the military, yet it has been mainly limited to facial recognition systems until recently. The US military has been experimenting with a variety of new AI systems.

AI drones

An AI drone can be defined as a drone that incorporates some kind of artificial intelligence, whether for route planning, facial recognition or target acquisition. Whilst the idea of AI drones is predominantly associated with the Russo-Ukraine War, it is far from limited to that conflict. The US military has been funding a programme called Loyal Wingman as early as 2023, which is focused on developing an AI fighter jet to function as a wingman to existing pilots (Karas, 2017). Furthermore, the US Navy is also developing AI ships known as Sea Hunters (Ehrlich, 2017), which are designed to run simple patrols around US waters.

Whilst these projects contribute to the idea of a Just War by starting to disassemble the idea of human soldiers and by turning the battles over life into battles of attrition and resources, this only becomes more just if both sides have these weapons in equal deployment. Furthermore, these systems have been used for scouting as well as allowing the surrender of soldiers, as in soldiers surrendering to a drone. There is also the element of better target acquisition by the AI system, yet this should require human oversight to avoid mistakes. A further use of AI is the idea of automated convoys to resupply conventional forces as well as deliver medical supplies. However, the darker side of this is the use of drones and drone swarms for kamikaze and suicide tactics.

Furthermore, it must be ensured that the AI drones target the correct infrastructure as well as factor in various mistakes that the system could make. This is particularly prevalent due to the extensive deployment of AI drones by Russia and Israel who have explicitly targeted civilians and civilian infrastructure. So, whilst these AI systems do save lives, historically, they mainly preserve the lives of the militaries deploying them. Furthermore, the deployment of these systems against a side with similar capabilities is only likely to prolong the warfare, as both sides will likely produce as many as they can without much consideration or care. This is due to people being more important than resources and once these resources are depleted, it is likely that the war will regress to a more conventional style. Hence, it is debatable how much AI drones will contribute to the idea of a Just War. Also, there is the concern that one side will deploy AI systems more readily than the other or have no regard for the proportionality of the force involved, meaning that it becomes a more one-sided war.

Finally, there is the issue of accountability; more specifically, who has it? The military that deploys it? The contractor that deploys it? The general who signs off on the drone deployment? The politicians that supported the bill through Parliament? When turning to the drone strikes carried out in the Middle East against the Taliban, it is the military, the drone operator and the personnel who planned the operation on whom accountability falls. Yet when AI reaches a point where it can pull the trigger with little to no human oversight, then accountability may fall to the developer of the system.

The AI problem with robotic soldiers

There is no doubt as to the effectiveness of AI drones, yet when turning to the idea of full robotic soldiers, there come the same issues associated with drones, including accountability. Yet whilst drones are often directed straight to targets and are as simple as fire and forget, when AI turns to robotic soldiers, new issues emerge.

One of the early issues is humans using very unorthodox tactics; for example, the Defense Advanced Research Projects Agency (DARPA) tested a small robot prototype to drive around with a marine squad for a few weeks. The idea is to test its performance of target spotting. Despite the great success, on the final day, the engineers tasked the marines with stealthily approaching the robot and touching it without being spotted. The first two marines simply somersaulted towards the robot, the third marine covered themselves in fir branches and walked in a tree-like fashion towards the robot. The final two marines located a large cardboard box, placed it over themselves and approached the robot. All succeeded in touching the robot without being caught (Scharre, 2023). So, whilst AI can be very effective, the poor identification could cost lives, whether the AI misses quite a clear threat, as in the case above or whether it mistakes something innocuous as a threat, possibly causing friendly fire or civilian casualties.

Despite these issues, AI soldiers could be utilised for more dangerous operations such as mine sweeping and clearing, as well as guarding convoys. The latter of which is likely only to have definite hostiles attacking it and therefore likely to be safer for everyone involved if AI soldiers are deployed. However, the issue here then becomes what classes as an attack, and at what point can an AI soldier engage? A rock thrown by a teenager towards an AI soldier is a provocation, yet not one which permits a soldier to open fire. Furthermore, that when applied to an AI soldier would classify as property destruction rather than assault, which in this situation would be inherently less violent. This becomes even more interesting when you apply traditional ideas of escalation, as the bombing of a battalion of human soldiers would be a tragedy, whereas the bombing of AI soldiers would merely be a slight concern. Therefore, the escalation would likely be much slower, or the damage would need to be catastrophic to meet a similar level of escalation. It is here where war is almost gamified, as there is little shock value with resource destruction due to the fact that more can be produced and rolled straight from the assembly line to the battlefield. There is also the issue of open-source software also being perverted for use in warfare. Yet one potential solution is developers writing in their GitHub pages that their repository cannot be used in warfare (Chan, 2019), yet this is far from a good prevention method. Indeed, there have been further attacks on the nature of restricting open-source software for specific tasks (Legal500, 2023).

Below, Nigel Crook offers his perspective on some of the challenges of robotics.

EXPERT REFLECTION

Nigel Crook, *Professor of Artificial Intelligence and Robotics at Oxford Brookes University, UK*

ARTIFICIAL LOVE?

I am frequently asked if it is harmful for people to develop romantic relationships with AI agents. This is rapidly becoming an area of major concern in AI ethics, where chatbot technology is facilitating increasingly realistic human-like interactions that are driven by a combination of human need for companionship and commercial interest. But is falling in love with an AI agent likely to be harmful and what are the ethical issues that surround this?

Healthy human romantic relationships are often characterised by mutual care, respect and affection, with an element of "give-and-take", and are nurtured through shared experiences and moments of being present to each other. Whilst AI agents and chatbots can be helpful in many ways, there is a real danger that they can give the impression of being all these things to an uninformed human user, leading them to believe that the agent can enter a romantic relationship with them.

Many of the dangers of AI-human relationships stem from a tendency on the part of users to forget or overlook the fact that these systems are sophisticated simulations of human behaviours and responses and that the words generated by an AI chatbot, for example, do not carry the same authenticity as when those words are spoken by a human. AI-based agents are not conscious, have not experienced the world in the manner that humans do and do not understand the human condition. Their outputs are generated by statistical processes rather than by deeply held convictions or moral principles, and they are not capable of experiencing emotion of any kind. On top of all of this, AI has no capacity to choose to enter into a relationship of any kind with a human.

Nevertheless, there are increasing numbers of people who are open to forming romantic relationships with AI agents. A recent Institute for

Family Studies/YouGov survey found that 1 in four young adults think that AI partners could replace romantic relationships with people.

There is recent evidence that young adults may be particularly susceptible to harm through their interactions with AI-powered chatbots with the tragic story of 14-year-old Sewell Setzer who took his own life following an engagement with a personalised chatbot. There is a desperate need for clear, ethical guidelines and policies about how people can safely relate to this technology. Above all, users of these systems should be aware in broad terms of how these technologies work and of how we can interpret their outputs and relate to them well as simulations of human beings.

Using AI morally in the context of mental health

A more moral use of AI is the utilisation of AI tools for mental health. For example, running emails through ChatGPT to ensure that the right tone is conveyed or an explanation is simplified. This improves a lot of internet accessibility for the elderly as well as those with autism, ADHD and those with other communication difficulties. However, despite this morality, there needs to be some limitation on this to ensure that people do not become over-reliant on this due to the AI overassuming or hallucinating. This becomes further exacerbated by the fact that humans can introduce their own self-bias into their prompts, leading to the AI being more likely to side with the user rather than condemning their action.

An interesting consideration in the article by Minerva & Giubilini (2023) discusses the idea that people are more likely to open up to a virtual interviewer since those users with PTSD or who are introverted are more reluctant to open up to another human. Furthermore, it also bypasses the societal stigma around mental health, as there is no human involved, only the AI, which will provide no judgement. Given that ChatGPT and AI can be accessed via a smartphone app, this does allow anyone to download an AI therapist, which further reduces the societal stigma, as no one has to know the person is receiving therapy. Furthermore, having an AI therapist app does mean greater accessibility and also the potential to reduce strain on mental health services such as Child and Adolescent Mental Health Services (CAMHS). Furthermore, providing patient data to the AI could be beneficial, as it could learn to provide better diagnosis for specific conditions such as ADHD or autism, since one of the main issues is getting a diagnosis due to healthcare systems being overloaded or doctor error. Overall, the paper concludes that AI in most scenarios will be beneficial to humanity yet may not be cost -effective. In the scenario where AI is dismissed, it is for the reasoning that AI had not yet surpassed humanity and that most of humanity may gain better results when interacting with an actual human. However, the first of these issues is likely to vanish in time due to AI advancement.

The paper by Koutsouleris et al (2022) further discusses AI implementation within the medical sector when regarding mental health. However, it focuses on more specific examples, such as the AIs that utilise the EMR-based database, an open-source database of electronic medical records. However, it does highlight that these records are made predominantly for individual patient care and have significant bias within them, therefore making the AI models biased or inaccurate. It further suggests that this bias can be aggravated to the point of prejudice and/or systemic racism within the models. The paper then argues for increased preciseness when collecting data in order to avoid this. Furthermore, the paper comments on a past approach using generative adverbial networks on COVID-19 data, that mitigated the bias and could be used in the future. Yet, because of the specific nature of treatments within these medical records and the more generalised route that the AI will have to utilise, the treatment suggested is likely to be compromised in a significant number of cases.

Furthermore, there would be notable behaviours and other variables that would be omitted from the EMR that would prove valuable to mental health diagnosis, further limiting the usefulness of AI systems based on this database. The paper goes on to discuss that data from smartphones as well as other wearables often lack information on certain events such as bouts of paranoia; thus, if this data is used, an imperfect model is constructed. Therefore, the paper argues for safeguards like better data collection techniques to mitigate this issue. However, the paper does go on to discuss the real-world implementations of mental health AI's, such as Facebook's Machine Learning tools to identify self-harm with Crisis Text Line implementing a similar system. The REACH VET Programme is known to utilise EMR as its system, which is designed to identify individuals at risk of suicide. The final real-world examples are the AI agent apps such as Woebot and Wysa that are designed to help manage anxiety and depression. Whilst currently limited, these apps present huge potential in the future, especially if they utilise more up-to-date AI models like Llama, DeepSeek or ChatGPT. However, when it comes to Woebot and Wysa, there are some prevalent issues of safety, such as these AI agents being unable to realise the dangers their users might be in and being unable to manage this or to contact the relevant authorities to ensure the safety of their users. Furthermore, the Crisis Text Line service received a major public backlash when they were caught selling the user data to a for-profit organisation (Lawler, 2022). Therefore, the question is, as Koutsouleris agrees, how can AI agents be made safe for the real world? How can researchers design AI that can follow the Hippocratic oath of do no harm? Therefore, a hybrid approach, with an AI supervised by a trained medical practitioner, does seem ideal. Yet this is likely to create a shortage of jobs in the future if the AI is perfected as the paper suggests. But a more worrying concern is when the opinion of an AI and a medical professional conflict, as who is to judge on who is correct? Especially if the AI has the ability to override the human or has greater respect due to the human's poor reputation, for example. This increased decision time and argument is likely to be detrimental to patient health and is further dangerous due to AI's inability to admit when it is wrong. Furthermore, AI hallucination may exacerbate less severe symptoms and therefore is likely to lead to more radical diagnoses, which could lead to increased mistakes as well as miscommunication. Therefore, the paper argues for a legal framework to be implemented for both the protection of both the AI and professionals so that they can feel comfortable in using AI.

Further moral use of AI and the deployment of it for the analysis of medical data such as cancerous cells. This can be done using traditional AI methods as well as Deep Learning, meaning that it is likely to use less power and resources than when traditional methods are utilised, reducing the environmental impact. However, when AI is used to enhance or infer data, this can lead to hallucination or errors, meaning that great care must be taken when developing these AI systems to ensure that they are accurate. Furthermore, these AI systems are often poor at predicting certain edge cases or anomalies, so to ensure the best results, medical personnel must not overly rely on the AI systems.

To conclude, moral uses of AI should include automating tasks that humanity finds boring and monotonous or providing only assistance to humanity for specific tasks that require it, rather than taking over the art and creative industry, which most of humanity enjoy doing. Yet when it comes to mental health, AI can certainly act as a limited medical practitioner. However, significant safeguards and legal frameworks must be implemented, as well as systems to mitigate bias and hallucinations. Furthermore, these systems should allow AIs some outside influence that goes beyond advising them to ring a mental support line, to allow them to contact authorities should they realise someone is in more significant danger.

Government stances

The government stance on AI has been rather lax in recent years, particularly in regard to social media, where members of parliament and congress have shown a clear lack of knowledge. This has been shown clearly in the congress hearings surrounding the banning of TikTok and the regulating of Meta.

As the paper by Radu (2021) describes, the main leaders in AI development aim to become world leaders in the technology, whilst EU member states and smaller countries more limited on AI technologies favour a more collaborative and international approach. This is likely due to these smaller countries falling behind technologically. A further limitation that is described is the lack of expert advice and thought process behind policy creation, as whilst some countries elected to do long consultations over their strategy, others simply delegated to one person (e.g. France) or to a group of experts (e.g. Finland). In these national strategies, only a few countries mention AI usage regarding defence, whilst seventy-five countries actively use AI for surveillance and facial recognition. Yet an even bigger concern is that few countries discuss the links between the public and private sectors of the AI industry as well as the general reluctance to strictly regulate AI. One of the more common strategies is the role of government in regulating the AI integration itself between sectors whilst remaining vague on the active role of government. This unfortunately does leave the door open for corporations to run away with the technology with little in the way of oversight, despite the general conclusion of these strategies ending in the creation of AI Councils or Committees to monitor AI adaptation and implementation. An even greater limitation of these councils or committees is the fact that although they are designed to focus on ethics, they are composed mainly of representatives of the private sector and academia with very little representation from rights groups or NGOs. This would indicate a greater focus on research and development rather than on whether specific research should be carried out. Overall, the paper concludes that whilst some good strides have been made towards AI regulation by governments, there will be significant hybridity between the ethical approach towards AI and the fast development of the technology to become the world leader in the technology. This is evident due to the vague guidelines laid out in these national strategies, yet it can be argued that these guidelines must remain more flexible to cope with growing advancements.

Cihon et al. (2021) consider the idea of corporate governance in artificial intelligence. One observation that the paper puts forth concerns Google's Project Maven, a military system that uses AI, and the influence of the workers as well as the media, which forced Google to back out of the project, which led to Palantir taking over (Daws, 2019). Yet the paper goes on to discuss that despite this success, the workers' influence may become lessened due to the increasing ease of replacing them with AI. Therefore, relying on workers to protest the use of AI may become useless in the future due to AI advancement and corporate greed. However, an interesting point that the paper proposes is regulatory diffusion when it applies to AI governance. Regulatory diffusion is the borrowing and implementing of the same or similar rules from another agency or body. The idea presented is that because corporations are multinational, they will likely follow the rules of the most stringent country, as that will ensure that they comply with all other countries' regulations. This will then later lead to them persuading other companies to adopt similar restrictions and potentially to lobbying governments to adopt the stricter restrictions in order to limit other companies to ensure fair competition and limit monopolies.

Conclusion

The use of AI will always be fraught with moral challenges, and great care must be taken to ensure moral data collection, moral AI development and deployment. In regard to moral data collection, a robust legal framework for collecting data with user and artist rights and consent; is required at the forefront to prevent the theft of data. When considering development and deployment, it becomes more challenging, as none can say what the AI will be developed into or if its use will be perverted after deployment. Whilst some argue that open-source software should be open to all uses, some protest that some uses should be prevented. When it comes to proprietary AI software, it is the responsibility of the corporation to ensure that the AI is developed ethically and morally. However, corporate greed often corrupts this; therefore, there must be independent investigation and verification of the implemented AI to ensure that moral and ethical principles are followed. Whilst governments around the world are setting up AI-focused councils to try and monitor the development of AI and its integration, Yet these are often limited in scope and lack much in terms of task forces or enforcement; therefore, governments must take steps to ensure that these councils and bodies have the relevant powers to intervene and supervise corporations effectively.

7

Ethics and AI

Introduction

Ethics is invariably seen in relation to the efficacy of a project or research study or, more broadly, as a general view about what counts as right and wrong in society. Yet ethics, particularly in AI, transcends every aspect of society, politics, higher education and religion, from virtual humans to plagiarism, manufacturing, self-driving cars and Large Language Models. This chapter begins by exploring what it means to be human and then reviewing the landscape of ethics in the 21st century. It then suggests practical responses to ethical concerns and explores ways of dealing with ethics and AI. The final section of this chapter explores the issues of ethics and AI in relation to the law and the impact of new legislation in 2025.

Understanding ethics and the nature of being human

In the context of AI, what it means to be an ethical human being has become a complex and somewhat troublesome question. Harari's (2015) arguments about what it means to be human is disturbing, such as the idea that human nature will be transformed in the 21st century, as intelligence becomes uncoupled from consciousness, and the suggestion that

> If you hear a scenario about the world in 2050 and it sounds like science fiction, it is probably wrong; but if you hear a scenario about the world in 2050 and it does not sound like science fiction, it is certainly wrong.
>
> *(Harari, 2017)*

Google and Amazon, among others, can process our behaviour to know what we want before we know it ourselves. Harari challenges us to consider whether indeed there is a next stage of evolution and asks fundamental questions, such as 'where do we go from here?' In a book about virtual humans in 2019, David Burden and Maggi Savin-Baden explored the notion of what it

DOI: 10.1201/9781003622949-8

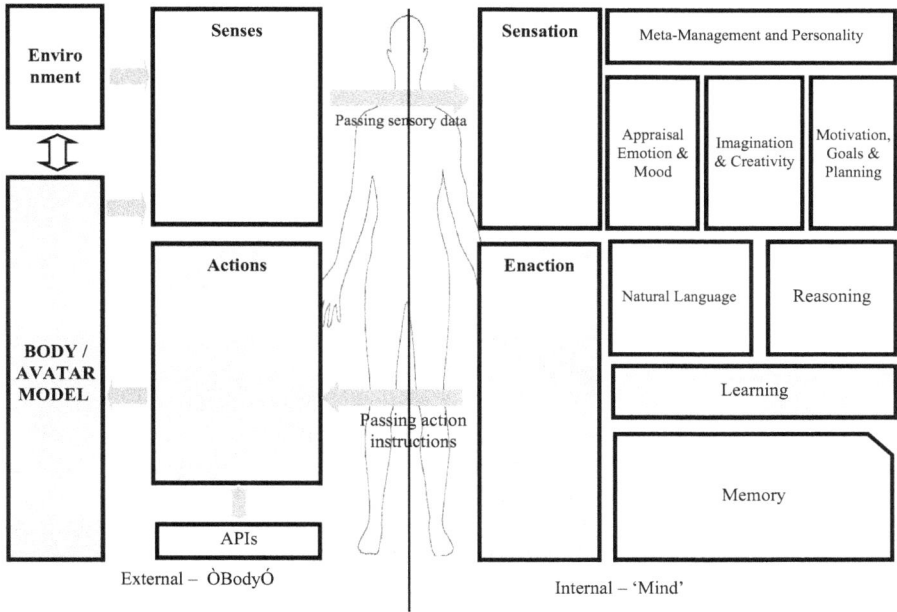

FIGURE 7.1
Elements of a virtual human (Burden & Savin-Baden, 2019, p. 4).

means to be human, arguing that a virtual human is fundamentally a computer programme: simply code and data that has been designed to give the illusion of being human. Figure 7.1 provides an overview of the possible elements of a virtual human, but not all of these elements would be present in every virtual human, and not every element needs to be present to the same degree.

The main elements are:

- A body, which may be a digital avatar or simply a microphone and speaker or text-chat interface
- A set of senses and the ability to detect sensations
- The ability to appraise sensation triggers and respond to them, including showing emotion and changing mood
- The ability to plan to achieve goals, ideally set by some internal motivation
- The ability to reason and problem-solve
- The ability to show imagination and creativity
- The ability to communicate in natural human language
- The ability to learn

- The ability to remember and access memories
- The ability to manage all the above, which may be reflected in a personality
- The ability to enact its decisions through taking actions with its 'body'
- Application Programming Interfaces (APIs) to other systems
- An environment in which to exist and interact with

(Burden & Savin-Baden, 2019, pp. 4–5)

The importance of understanding what it means to be human is important because in ethical debates about AI, virtual humans are designed using human decision criteria, and therefore, ethical behaviour needs to be 'designed-in' to virtual humans. We see being human as having an ethical compass, a sense of our own sentience and impact on the world. Being human for us also invokes the morals of 'do no harm' and 'love thy neighbour'. Thus, it is important to fight against injustice and selfish capitalism. What we mean by selfish capitalism is the idea that capitalism is less about the free market and more about a focus on selfish personal or national gain. It is also important to stand against modern slavery and any kind of exploitation, whether from people, companies or covert AI practices that undermine what it means to be human.

Brown (2021) argues that to be human is to be embodied and rational, but he also argues for the need for us to examine how humans become moral beings. He draws on the work of MacIntyre (2016) to reconstruct an account of how we become moral persons through participation in communities and create narratives that help construct the self and address both human dependency and autonomy. The difficulty then with developing adequate morals for humans, never mind virtual humans, is that responses to moral dilemmas vary greatly since they emerge from cultural and moral traditions. Thus, in the context of the range of virtual humans that are being developed, designing appropriate and effective ethical standards is complex and far-reaching. Currently, there is not a single global jurisdiction, and the laws that apply in one country or state may not necessarily apply elsewhere, which means that the use and abuse of virtual humans across the world, particularly in terms of those who create virtual humans for illegal purposes for use in a different country from their own, are unlikely to be prosecuted. The recent European Artificial Intelligence Act in 2024 (European Parliament & Council of the European Union, 2024) aimed to regulate AI systems, eliminate obstacles to trade and protect society against the adverse effects of AI. To date the UK has not signed up to this due to it no longer being in the European Union, and there is currently much debate about its values, as will be discussed later in the chapter. Thus, in the context of AI ethics, it is important to consider different ethical theories and options. Claddagh NicLochlainn offers her perspective on some of these issues.

EXPERT REFLECTION

Claddagh NicLochlainn, *Parliamentary lead at the Institute for the Future of Work, UK*

RULES FOR MACHINES, RIGHTS FOR PEOPLE *THE ETHICAL FRONTIERS OF AI GOVERNANCE*

As artificial intelligence becomes embedded across every domain of human activity - from education and policing to hiring, healthcare and warfare - the ethical questions it raises can no longer be treated as an afterthought. The technologies we design reflect the values we encode, whether intentional or not. Ethics, then, is not an external constraint on AI but a core design principle, shaping what these systems are for, who they serve and what futures they enable.

At the heart of ethical concerns about AI is the question of power. Who builds AI, who benefits and who bears the risk? Automated systems now make decisions about immigration, benefits and job applications, yet many operate with little transparency, recourse, or democratic oversight. This raises urgent questions about autonomy, fairness and accountability. In some cases, these systems replicate historical biases. In others, they obscure responsibility altogether, making it difficult to contest harm or demand redress.

As AI becomes more complex and autonomous, the stakes grow. Militarised uses of AI, such as autonomous weapons or predictive surveillance, raise profound questions about human judgement and the ethics of delegating life-and-death decisions to machines. In other ways, they force us to confront questions about human decision making and ethics more broadly – what do the values we encode in these systems tell us about the values that have been driving our own human decision making for so long?

The last decade has seen a proliferation of AI ethics principles, from the Asilomar AI Principles to the OECD and UNESCO frameworks. Common themes include transparency, accountability, fairness, privacy and human-centredness. Yet putting these principles into practice remains a challenge.

Some jurisdictions have taken the lead. The European Union's AI Act, introduced in 2025, categorises AI systems by risk and imposes legal obligations accordingly, banning certain uses outright, such as manipulative biometric surveillance; and requiring robust human oversight in high-risk sectors like health and education. The UK's evolving regulatory approach, while more sector-led, is beginning to incorporate ethical risk assessments and algorithmic audit requirements, particularly in public sector use.

Research institutions, too, are responding. Some universities have adopted ethical review boards specifically for AI development, treating these technologies with the same care as clinical trials or human subject research. Others are embedding ethical reasoning into computer science curricula, helping future developers consider the societal consequences of their work.

Ethics in AI is not a box to tick or a risk to manage; it is a fundamental question about the kind of society we are building. As Large Language Models shape communication, recommendation algorithms shape belief and synthetic content blurs reality, we are confronted with foundational questions about what it means to be human, to be free and to be responsible.

Good ethical practice in AI does not promise perfect outcomes. But it insists on scrutiny, participation and care. It demands that we ask not just what AI can do, but what it should do and who gets to decide.

The next section of this chapter explores traditional stances towards ethics in relation to AI but then argues that modest ethics should be the central plank of using AI in ethical ways. Table 7.1 brings together both traditional ethical theories and also newer stances towards ethics such as modest ethics. It delineates eight different ethical approaches, outlines their possible stances towards the postdigital and suggests probable practical challenges. This table summarises and suggests that there are difficulties with some of the more traditional ethical stances. Furthermore, it is argued here that modest ethics offers the most realistic and unproblematic means of aligning ethics in the context of the challenges of AI. We suggest the issues of ethics in AI need to transcend earlier ethical theories and examine truths in ways that are pragmatic and explore moral problems in context.

A different stance on ethics

In the rapidly changing world of AI, in all of its diverse forms, we believe a new stance towards ethics is required; that of modest ethics. Modest ethics embraces a deep interconnection between ethics and materiality, leading to a new kind of worldly ethics, which is more-than-human in its scope. The idea of modest ethics can be aligned with the idea of the postdigital (for example Jandrić et al., 2018), since both prompt a rupture in our existing theories. The modest ethics approach rejects overarching ethical frameworks and their limited respect for the empirical complexities of real-world scenarios and the power and agency that such frameworks ascribe to human rationality. As Reader and Savin-Baden have argued:

TABLE 7.1

Ethical theories

Form of ethics	Definition	Key characteristics	Perspectives on AI	Practical challenges
Deontological	The focus is on the rightness or wrongness of the actions themselves rather than the consequences.	Focus on duty and obligation.	AI is a difficult concept to accept as, in general, it does not deal with right and wrong.	Difficulty in accepting the idea of postdigital ethics when there is an assumption that there are clear boundaries.
Utilitarian	The assumption that ethics can be governed by deciding what is right and what is wrong.	Belief in the greatest good for humanity.	AI is a troublesome concept as it has few boundaries and is always changing.	Notion of outcomes and what counts as best is complex in the postdigital realm.
Feminist	The belief that traditional ethical standpoints have theorising that has undervalued women's moral experience.	Stance toward female empowerment and against oppression.	AI can help to challenge and interrupt toxic misogyny.	Gender-based oppression in postdigital spaces is hard to counter.
Marxist	This is based on the idea that morality is a historical product, and people are conditioned by this.	Stance against elite power and economic exploitation.	AI is seen as interrupting elite narratives and surveillance practices which is a positive position.	Surveillance capitalism is largely ignored globally.
Situational	The situation itself should be evaluated ethically rather than being judged by predefined moral standards.	Rejection of prescriptive rules and decisions.	AI is valued because the focus is on context rather than judgement.	Circumstances difficult to judge and forgiveness often challenging.
Discourse	Ethics and norms are created through discourse.	Valuing discourse as a means of reaching understanding and agreement.	The belief here is that discourse can help to develop understanding of postdigital ethics.	Discourse across cultures poses challenges in postdigital ethics.

(Continued)

TABLE 7.1 *(Continued)*

Form of ethics	Definition	Key characteristics	Perspectives on AI	Practical challenges
Virtue ethics	Virtue ethics treat the concept of moral virtue as central to ethics.	Actions are those that embody virtuous character traits, such as courage, loyalty, or wisdom.	The importance here is in acting virtuously in digital spaces by practising inclusivity.	The assumption that virtuous person is able to determine and reason what counts as right action.
Modest ethics	Modest ethics affirms what a body can do, with what degrees of power, movement or rest.	Facing what we cannot control Enhancing life	This approach would see AI as offering possibilities for life enhancement, whilst also arguing that the (misplaced) power of AI needs to be challenged continually.	The hope to counter the speed of the digital The optimism in being able to disrupt established binaries of human and nonhuman capabilities.

Ethics, along with both science and religion, need to exercise a degree of modesty. Claims to establish truth in some exclusive manner, and therefore at the cost of an open engagement with other contributions, need to be tempered by the recognition that no one approach has the monopoly of truth.

(Reader & Savin-Baden, 2020, p. 300)

There is a tendency to construct ethical frameworks in order to cope with such complexity and produce criteria by which we can evaluate human action and make decisions about the right way forward. Modest ethics encourages us to become more comfortable with that which we cannot control and which is constantly moving beyond us. Modest ethics may demand that we move more slowly and carefully, rather than dashing ahead in ways that our culture and digital technology encourage us to do. Reader and Evans suggest that modest ethics includes the ethics of information and the ethics of digital techno-cultures and is modest with regards to:

- their rejection of overarching ethical frameworks and their respect for the empirical complexities of real-world scenarios
- the amount of power and agency that they ascribe to human rationality
- their focus on the deep interconnections between ethics and materiality

(Reader & Evans, 2019, p. 27)

Recovering a language of virtue and modesty will also help us to deal with issues such as ethics in relation to Machine Learning, as well as dupery and deception.

Ethics and Machine Learning

Ethical concerns in the field of Machine Learning can be considered in relation to robotics ethics, machine morality and intimacy ethics. For example:

- Robot ethics needs to examine ethical questions about how humans should design, deploy and treat robots.
- Machine morality introduces questions about the moral capacities a robot or virtual human should have and how these might be implemented.
- Intimacy ethics explores how engaging with virtual humans can offer people opportunities to connect with something emotionally and feel supported, even loved, without the need to reciprocate.

Jandrić (2019) points out that biases are not just carried through cultures and people creating technology; biases are built into technology in ways that are unpredictable. This in turn raises further questions about the use of Machine Learning, which is essentially designed to enable computers to act without being programmed. Thus, as aforementioned, some of the ethical issues are that whilst ethical behaviour needs to be 'designed-in' to virtual humans and robots, designing appropriate and effective ethics standards is complex and far-reaching. For example, if autonomous combat robots or AIs are deployed by the military, whose fault is it if they mistakenly attack or fail to distinguish correct targets? In online warfare and wargames, the findings of a literature review (Savin-Baden, 2025) focus on the role of the game designers and developers regarding Just War, arguing that they must ensure that the players have the option to conduct Just War. However, the approach generally taken is that players, regardless of game features, will find ways to fight both justly and unjustly such as the WoW players assaulting the virtual funeral as Losh (2009) describes. Furthermore, Guanio-Uluru (2016) discusses that wargaming, particularly in fiction such as Ender's Game and The Hunger Games, is inherently unjust with diabolical tactics used. However, the conclusion is that peaceful games are generally more popular than wargames, implying that most gamers do strive for peace. The concerns about Machine Learning, drones and gaming, whilst complex, are being addressed through research and literature reviews. However, what is less evident is the level of dupery that is occurring.

Dupery and deception

The idea of dupery implies that we are tricked in some way, hoodwinked in ways that make us feel the mistake is ours and that we are rather stupid in being so easily taken in. This is something that occurs often in social media when people are misrepresented; their photographs, words and views are altered, revised and manipulated in unscrupulous ways. In the 21st century, the most common form of acceptable dupery is that of magic, largely because we know we are being deceived, and we are fascinated by it. Sinclair explains that what is central to dupery is the use of emotion to deceive, which occurs through manipulation, distraction and even laughter (Sinclair, 2021). Furthermore, she explains that different types of duperies can have different effects and that the common tactics used to manipulate us are misdirection and distraction. She cites a poignant example of where deception was added to a simulation resulting in the death of a mannequin where a senior clinician entered the simulation and asked for the wrong medication to be given. A clear case of planned misdirection. Calhoun et al. describe what occurred in this simulation and argue that there may have been concerns about the stress that such dupery has on students (Calhoun et al., 2015). What such practices do indicate is the need to ensure students are properly debriefed after such deception, something that is also argued for and undertaken in psychology experiments where students are deceived. Whilst the instances she describes do have hidden emotional practices, other practices have more of a sense of trickery such as deception and fake news.

When attending a magic show, people expect to be tricked, but possibly not in an online seminar. With magic it is clear that deception occurs through distraction and changing perception of what is occurring. To some extent this is also the case when being deceived in higher education. Few people anticipate deception in a learning environment, and therefore, when they are tricked this can result in anger and frustration. Emotional and cognitive disjunction of the deception undertaken by Taras and Steel demonstrates this (Taras & Steel, 2007). The students were told by Taras that their usual lecturer was suspended in a sympathetic way and explained to the students there would be some changes in the assessment. The students complained, so Taras provided a signup sheet for them to join a union and take their complaints further, then watched as the students came together as a group to take action. The students were later debriefed about the deception within the lecture, thereby illustrating to them how easy it was to violate trust and psychological contracts. A further example of this was a study that included the use of pedagogical agents within an online problem-based (PBL) seminar series (Savin-Baden et al., 2016). A pedagogical agent (or chatbot) is a software application that can provide a human-like interaction using a natural language interface. Examples of these are Siri, Cortana, Alexa, or the virtual online assistants found on some websites, such as Anna on the Ikea website. The passive detection test is where participants are not primed to the potential presence of a pedagogical agent within online PBL. The active detection test is where participants are primed to the potential presence of a pedagogical agent. The study used PBL online so as to give a focus for discussions and participation without creating too much artificiality. Students, in the main, did not detect the agent, and the ways in which students positioned the agent tended to influence the interaction between them. Thus, those who assumed the agent was an incompetent student ignored him and those who believed he had language difficulties felt sorry for him. Students' positioning of the agent – as shy, arrogant and confused – was largely a mechanism used for rationalising their feelings that something was awry or uncanny.

Bullshit and lies

The term 'bull' dates from the 17th century, meaning nonsense; the world 'bullshit' gained popularity in the 1900s and was probably first used by the poet T.S. Eliot in his poem The Triumph of Bullshit in 1910 – even though the word does not appear in the poem itself. Current usage is seen as slang and still refers to something being nonsense but also foolish, insolent talk. Mackenzie and Bhatt (2020) distinguish between a lie and bullshit as follows:

- bullshit, unlike lying, may not undermine trust, but instead plays with or exaggerates the truth and tends to be overly dramatic
- a lie is a statement made that is known to be false, with the intention of misleading or deceiving for advantage

Thus, people may not realise they are being lied to, and lies tend to destroy trust – unless they are just polite untruths to avoid offence such as 'that was a lovely present'. However, invariably most people recognise bullshit, particularly that occurring on social media and even more so with President Trump of the USA. The difficulty with the latter kind of bullshit is that it is difficult to know if there are hidden truths behind it, such as the argument in 2025 that the US will buy Greenland or Gaza. So such suggestions may or may not be bullshit, but they are used as deflections in order to shift the attention away from other practices or laws President Trump wants to hide from the public gaze. Thus, this kind of bullshit is probably more dangerous than lies because in 2025 news is being screened and manipulated by the Republican Party so that truths about immigration, the treatment of women and climate change are hidden from the gaze of the American public.

Fake news and fake representation

MacKenzie and Bhatt (2020) suggest that fake news may not be intentionally misleading since often the intention is to satirise real events. However, 'fake news' was a term that was popularised during the 2016 US elections in order to describe both inaccurate news and staged news shows. It is an important concern for students to be aware of when learning in higher education. Otrel-Cass and Fasching (2021) explored young people's encounters with fake news and found that in order to enable them to understand the potential instances of dupery, teaching in this area needed to be personalised to reflect students' experiences of online and fake news. Özdan (2021) explored whether legal remedies against the circulation and publication of fake news were in fact compatible with international human rights law. His findings indicated that when dealing with fake news, policy makers should adopt reason and should expect conformity which upholds the principles of international human rights law. Furthermore, he suggests that the policy of censorship is not the solution, but instead, media literacy programmes and transparency policies need to be adopted (p. 93).

In the 2020s, the growth of Large Language Models and the use of AI to create and change images is a further concern. A recent BBC news story was told of the journalist and newsreader Naga Munchetty whose name and image were used by scammers to attempt to extort money by creating and using a fake BBC news site. She explains

> The fake article about me suggested I had been detained by the government following a "controversial" interview on ITV's This Morning, where I allegedly gave details about a "lucrative loophole" to make money. It was made to look like a BBC News article, complete with logo and branding, and it contained links to a scam cyber trading website, which has now been taken down after my production team reported it to the BBC legal team.

(Munchetty, 2025)

Such misuse of images has happened to other BBC presenters, and in doing so, this has sought to undermine both the BBC and its presenters.

A study by Reicho and Otrel-Cass (2024) explored how visual information affected students' evaluation of online information. Using video ethnography, analysing field notes and undertaking semi-structured interviews with secondary school student pupils, the study identified two levels of trust. The two levels were proximity to human actors and access to digital information, as illustrated in Figure 7.2.

What is important about this study is not only its findings but also the way in which it raised young people's awareness of fake news. However, the study also indicated that the methodology they adopted to teach the students did not alert the pupils to the dangers of trusting free tools and Large Language Models.

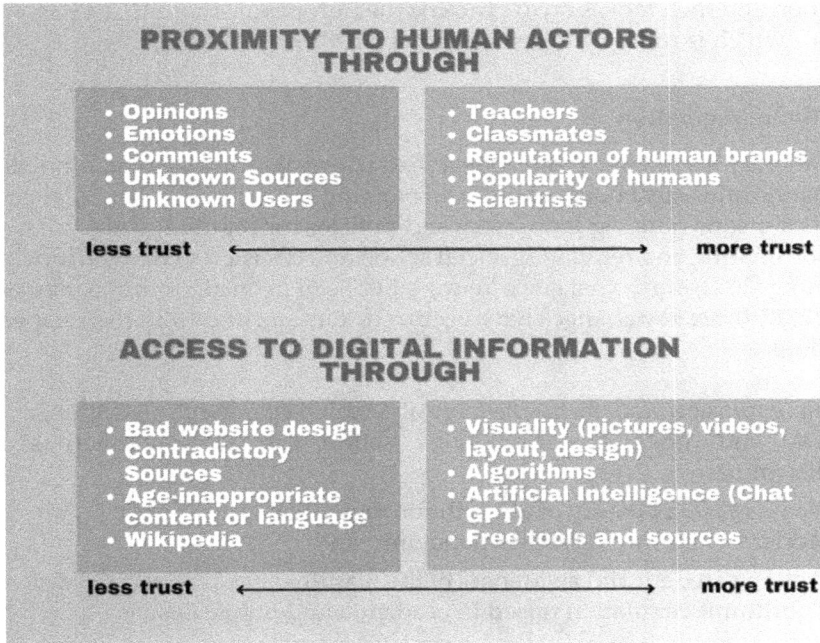

PROXIMITY TO HUMAN ACTORS THROUGH

- Opinions
- Emotions
- Comments
- Unknown Sources
- Unknown Users

- Teachers
- Classmates
- Reputation of human brands
- Popularity of humans
- Scientists

less trust ←————————————→ more trust

ACCESS TO DIGITAL INFORMATION THROUGH

- Bad website design
- Contradictory Sources
- Age-inappropriate content or language
- Wikipedia

- Visuality (pictures, videos, layout, design)
- Algorithms
- Artificial Intelligence (Chat GPT)
- Free tools and sources

less trust ←————————————→ more trust

FIGURE 7.2
Trust in human actors and digital information by levels of proximity and accessibility (Reicho & Otrel-Cass, 2024, p. 13) (Licensed under CC BY 4.0).

The stealing and reuse of data and images, along with trusting free tools and LLMs is invariably something few people consider in terms of the impact on people's revenues (such as the loss of income to artists whose original work is marginalised by the use of AI images) and the scraping of data from people's Facebook and Instagram accounts. Thus, it is clear, like the secondary school pupils, that everyone needs to be more aware and have a greater understanding of how to deal with identity issues online and be encouraged to manage and control the information contained in our own online networks.

Law, ethics and AI

The legal landscape worldwide is constantly shifting, and therefore, it is vital to be aware of current debates and laws. Discussions about law and AI, particularly the creation of the group to develop the European Union AI Act, have resulted in debates concerning decisions about AI and how such decisions are made. This section begins by exploring the concept of algorithmic regulation and then moves on to examine the European Union AI Act passed in 2024 – which is unlikely to be enacted until 2026.

Algorithmic regulation

As mentioned in Chapter 1, the widespread use of algorithms to manipulate information and affect political life is an ongoing concern. Algorithmic regulation is a means of using laws, rules and policies for regulating algorithms and coordinating and regulating social action and decision making. Ulbricht and Yeung (2022) argue that since Yeung's publication on algorithmic regulation in 2017, three key changes have occurred, summarised from their paper, as follows:

1. the explosion and diversification of scholarship about algorithms, data and digital transformations within and across numerous disciplines;
2. the increasing take-up of algorithmic regulation in the public sector, accelerated by the COVID-19 pandemic; and
3. growing anxiety and awareness of the manifold hazards and risks of algorithmic regulation raised in academia and public debate.

In 2025 the EU and UK are encouraging the use of AI for algorithmic regulation, for example, to boost law enforcement. Smuha (2024) suggests that such enforcement can erode the rule of law and argues that outsourcing important administrative decisions to algorithmic systems undermines the principles of democracy. Public authorities are increasingly turning to algorithmic regulation, or the use of algorithmic systems to apply and enforce the law. Their adoption of algorithmic regulation tends to be motivated by the desire to improve public services and citizens' rights. The difficulty is, as Smuha (2024) points out, that as laws are changed from text to code changes occur to the law; thus, it is undermined by algorithmic regulation. Changing law to code may at first seem to be a technical exercise, but it is not, for example, the notion of refusing to cooperate is very complicated and is open to interpretation, yet it may seem straightforward in an algorithmic system. Such issues tend to be overlooked when using algorithmic regulation to manage policy and law, and in particular, she suggests that the EU's digital agenda is misaligned with its aim to protect the rule of law.

The European Union's AI Act

The European Union's AI Act was adopted by the European Union in spring 2024, but it was not adopted by the UK since it is not part of the EU. The UK does plan to develop its own legislation. However, in February 2025, the UK and US refused to sign an AI declaration at the Paris summit that was supported by France, China, India, Japan, Australia and Canada. The UK felt the declaration had not gone far enough in addressing national security on the global governance of AI (Milmo & Courea, 2025). It is also probably safe to assume the US did not sign, as it does not want any AI regulation at all. The EU AI Act seeks to regulate AI systems and protect the EU's core values against AI (Smuha & Yeung, 2024). A feature of the Act is that it seeks to harmonise the legislation of member states and reduce obstacles to AI trade. The Act has also adopted what it terms 'a risk-based approach', the idea that graded tiers are used to classify AI systems. In practice, this means five tiers:

1. Systems that pose an 'unacceptable' risk are prohibited;
2. Systems deemed to pose a 'high-risk' are subjected to requirements akin to those listed in the Ethics Guidelines;
3. GPAI models are subjected to obligations that primarily focus on transparency, intellectual property protection and the mitigation of 'systemic risks';
4. Systems posing a limited risk must meet specified transparency requirements;
5. Systems that are not considered as posing significant risks do not attract new legal requirements.

(Smuha & Yeung, 2024, p. 11)

Additionally, the Act prohibits unacceptable risk, which relates to the exploitation and manipulation of human behaviour and lists a range of what it deems to be high-risk AI systems. The latter ranges from biometric systems to education and employment, justice and law enforcement. The Act also introduces specific provisions for general AI models such as Large Language Models. Whilst the Act offers detailed coverage of many of the challenges of AI, it is clearly more aspirational than practical. Furthermore, it is not yet clear how the Act will be put into operation, and with the rapidity and development of AI systems such as DeepSeek (an LLM developed by a Chinese artificial intelligence company to rival ChatGPT, launched in early 2025 and funded by the Chinese hedge fund High-Flyer), it is difficult to know how any legal framework will be able to remain up-to-date amidst the speed of change.

Responsibility and values

The complexity of dealing with ethics and AI not only relates to the speed of technological change but also to the shifts that are being seen in values and notions of responsibility. The lack of personal regulation of speech and behaviour by business, political and societal leaders in the public domain has resulted in a public perspective that bullshit, scapegoating and lying are acceptable practices. Furthermore, as was discussed in both Chapters 1 and 6 surveillance, bias and privacy are some of the many moral concerns that need to be examined constantly as AI and AI ethics continue to develop.

The notion of responsibility has been debated over many years and has been debated in relation to whether human beings are deemed to have free will or not. In the main, the overarching understanding of responsibility is related to making judgements about whether a person is morally responsible for their behaviour, as well as making decisions about our own and other people's responsibility for their actions and the consequences of those actions.

One of the difficulties with the increasing use of AI for everything is that understandings of moral responsibility are changing. Rethinking responsibility and values in the context of AI is complex, because it demands a sense of ownership – a sense of being responsible for one's action in the face of AI. Whilst it is not possible to hold individuals accountable for mistakes with an AI system, it is possible to use AI in responsible ways. For example, people can be responsible for the extent to which they use AI to cheat in an assignment or use it so much that it has a high impact on climate change. What is central to understanding AI ethics in practical ways is being able to take an informed stance on issues, such as the following questions we asked a UK group of theologians recently about whether it is ethical to:

- Use ChatGPT to change my references from Harvard to APA
- Download a video sermon on Malachi and then put it through AI software to use it in a sermon and pass it off as one's own work
- Using Sensay software to turn a book chapter into a podcast
- Ask ChatGPT to create prayers for me to use on Racial Justice Sunday
- Use a webcam in church children's groups for safeguarding purposes
- Adopt facial recognition software for use in church and mission buildings

When discussing issues such as these, many people have noted that they are unaware that sharing their own published books and articles with LLMs such as DeepSeek or ChatGPT results in them losing their own copyright, or indeed that data used by these LLMs are predominantly stolen. Thus central to AI ethics is choice and moral responsibility, the idea that we can choose to have ethical values and act in ethical and responsible ways.

Conclusion

The difficulty with the area of ethics and AI is that the AI landscape is continually on the move. A further concern is that dupery and lying has increasingly become part of the political canon so that the values of the good society – the values of equality, democracy and sustainability are continually being eroded. The general worldwide acceptance of LLMs as 'a good thing' illustrates that there is little understanding or regard for the unethical data scraping practices used to produce them. We argue that there need to be better and more effective frameworks and laws that are actually possible to implement, but more important is the need to debate these concerns deeply and widely across social, political and education sectors worldwide, in the hope that laws will be developed that can be implemented effectively.

8

Realistic options

DOI: 10.1201/9781003622949-9

Introduction

This chapter presents some of the realistic options when it comes to responding to some of the challenges of AI. This chapter begins by presenting the serious mistakes some companies have made in publicising their increasing commitment to AI. The second section discusses the user of the AI systems themselves, such as the ignorance and negligent nature of users, as well as what the model user should do in response to the world becoming more AI-driven. The final section discusses AI poisoning techniques for use as defence against AI systems. The overarching argument of this chapter is that the perspective towards AI needs to change; from a corporate perspective, humans should be prioritised in terms of employment, and AI must be implemented in ways that are ethical. In terms of the users, people need to be more cautious about their level of use of AI, as well as ensure that their use of AI is ethical.

The corporate response

With the rise of AI, institutions and companies have been eager to implement AI systems both to gain efficiency and to implement cost-saving measures. Yet instead of being able to achieve major successes, some corporations have faced massive criticism due to both job layoffs and the problems with the AI systems themselves. Therefore, these corporations suffer mainly negative publicity, whilst AI companies report soaring profits. An example of this is the NaNoWriMo scandal, presented below.

DOI: 10.1201/9781003622949-9

The bad – the fall of NaNoWriMo

The first disaster to be covered is the NaNoWriMo scandal. NaNoWriMo stands for National November Writing Month and is a non-profit organisation designed to encourage young writers to produce a manuscript during the month of November and then go on to publish it. In May 2024, NaNoWriMo underwent a mass restructuring due to a previous controversy, and in doing this, they announced their intention to encourage AI writing tools, as they had partnered with an AI writing company known as ProWritingAid (Figure 8.1). The intended impact of this was to recoup some of their lost funds, but NaNoWriMo only succeeded in drawing significant criticism.

Furthermore, they marketed these tools as a way to combat classism and ableism (Figure 8.2), arguing that not everyone had access to good education or writing resources. The latter point is rather ironic, given that NaNoWriMo prided themselves on providing the best resources for young writers.

However, these changes and explanatory statements did not have the effect that NaNoWriMo intended and promoted a huge backlash as well as a mass NaNoWriMo account deletion. Many writers viewed NaNoWriMo's introduction of AI writing tools as a way to cheat the system, since similar AI writing tools such as NovelAI, Red quill and ChatGPT can generate entire books from short prompts with only some editing needed from the user. Many users took to Tumblr to express their outrage. Instead of reversing their decision, NaNoWriMo elected to continue to defend their decision, leading to rather heated discussions between NaNoWriMo and their users on social media. Eventually the organisation realised that they were fighting a losing battle and backtracked, issuing official revised statements as seen in Figure 8.3 and deleting the ones shown in Figures 8.1 and 8.2. It was only due to users screenshotting the original statements that they were preserved.

This event proved that the writing community felt that writing was best left to humans rather than AI and that they would not take being replaced lying down. Furthermore, this illustrates the wider challenges of AI, which enable almost completely replacing their writing staff with AI tools, which means that more scandals like NaNoWriMo are likely to occur. Furthermore, the use of AI tools means a significant decrease in costs for corporations, which makes it very attractive. Yet when corporations alienate writers, there is no one to ensure the quality of the AI-produced work, to fact-check or to ensure mistakes made by the AI are corrected. Thus, if a corporation solely relies on AI for writing, it means that the sole writing voice of the company is an AI which is not fully reliable, which can lead to numerous concerns, for example, if the AI makes incorrect statements or states something that can be easily misinterpreted.

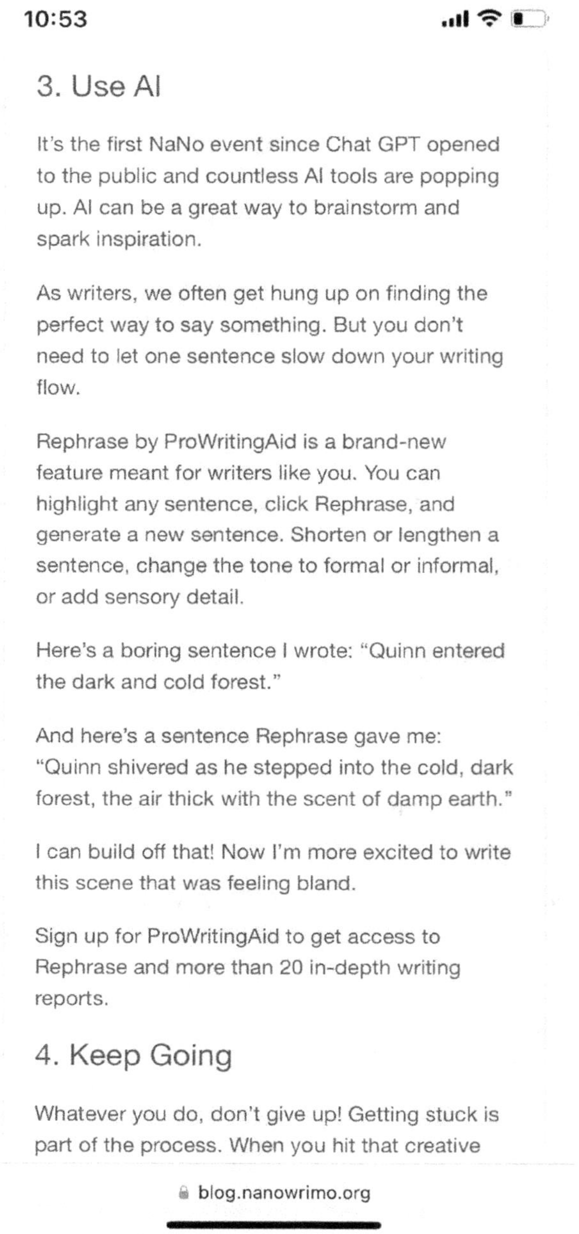

10:53   ..Il 📶 🔋

3. Use AI

It's the first NaNo event since Chat GPT opened to the public and countless AI tools are popping up. AI can be a great way to brainstorm and spark inspiration.

As writers, we often get hung up on finding the perfect way to say something. But you don't need to let one sentence slow down your writing flow.

Rephrase by ProWritingAid is a brand-new feature meant for writers like you. You can highlight any sentence, click Rephrase, and generate a new sentence. Shorten or lengthen a sentence, change the tone to formal or informal, or add sensory detail.

Here's a boring sentence I wrote: "Quinn entered the dark and cold forest."

And here's a sentence Rephrase gave me: "Quinn shivered as he stepped into the cold, dark forest, the air thick with the scent of damp earth."

I can build off that! Now I'm more excited to write this scene that was feeling bland.

Sign up for ProWritingAid to get access to Rephrase and more than 20 in-depth writing reports.

4. Keep Going

Whatever you do, don't give up! Getting stuck is part of the process. When you hit that creative

🔒 blog.nanowrimo.org

FIGURE 8.1
NaNoWriMo statement regarding ProWritingAid.

- **Classism.** Not all writers have the financial ability to hire humans to help at certain phases of their writing. For some writers, the decision to use AI is a practical, not an ideological, one. The financial ability to engage a human for feedback and review assumes a level of privilege that not all community members possess.
- **Ableism.** Not all brains have same abilities and not all writers function at the same level of education or proficiency in the language in which they are writing. Some brains and ability levels require outside help or accommodations to achieve certain goals. The notion that all writers "should" be able to perform certain functions independently or is a position that we disagree with wholeheartedly. There is a wealth of reasons why individuals can't "see" the issues in their writing without help.
- **General Access Issues.** All of these considerations exist within a larger system in which writers don't always have equal access to resources along the chain. For example, underrepresented minorities are less likely to be offered traditional publishing contracts, which places some, by default, into the indie author space, which inequitably creates upfront cost burdens that authors who do not suffer from systemic discrimination may have to incur.

FIGURE 8.2
NaNoWriMo statement on Ableism and Classism.

A Note to Our Community About our Comments on AI – September 2024

To the NaNoWriMo Community:

In early August, debates about AI on our social media channels became vitriolic. It was clear that the intimidation and harassment we witnessed were causing harm within our community of writers. The FAQs we crafted last week were written to curtail those behaviors. We wanted to send a clear signal that NaNoWriMo spaces would not be used to bully or delegitimize other writers. This was consistent with our May 2024 statement, which named a lack of civility in NaNoWriMo spaces as a longstanding concern.

Taking a position of neutrality was not an abandonment of writers' legitimate concerns about AI. It was an acknowledgment that NaNoWriMo can't maintain a civil, inclusive community if we allow selective intolerance. We absolutely believe that AI must be discussed and that its ethical use must be advocated-for. What we don't believe is that NaNoWriMo belongs at the forefront of that conversation. That debate should continue to thrive within the greater writing community as technologies continue to evolve.

Our Mission is about providing encouragement to writers and cheering them on as they progress toward their goals. That remains our primary focus. We apologize that our original statements lacked appropriate context and that our mistakes created distress. In the future, we will be more transparent about the issues that we are trying to address with any messaging we provide.

Finally, we recognize that some members of our community have other questions. We don't think we can address all of these in a single communication. Additional context here is that we are a very small team (including our Interim Executive Director, who is a volunteer). We want to take the time to read through your letters with the care, attention, and concern they deserve. Please expect more in the coming weeks.

In partnership,
The NaNoWriMo Team

FIGURE 8.3
NaNoWriMo September statement on AI.

The stupid – the owl's obliteration

The second issue to be covered is the downfall of Duolingo, a language learning app that was built on a brilliant marketing campaign by their mascot, the Duolingo owl. The app had 9.5 million paid subscribers by the end of 2024 (D'Souza, 2025) and even for those who did not download the app, most people knew of the owl and its unhinged behaviour both on the Duolingo app itself and on the Duolingo TikTok account. This behaviour appealed greatly to the younger generation, often dubbed Gen-Z, who made up most of the app's following and user base (Gagne, 2023). Unfortunately, the great success of this app was brought crashing to the ground by one foolish statement. In May 2025, Duolingo issued an email from their CEO announcing their intention to become an 'AI first' company (Figure 8.4) and to gradually lay off their contractors in favour of AI systems both to carry out their main business of translations and in ancillary sectors of their business such as recruitment and evaluations.

Duolingo also released 148 new courses written by Generative AI (Figure 8.5)

This came as a serious shock to their user base, and Duolingo faced huge repercussions. Despite their attempts to fight the backlash through memes and their typically unhinged conduct, their users refused to back down and responded by attacking Duolingo's TikTok account. After over ten thousand hate comments on all of their latest TikTok posts, Duolingo eventually deleted their main TikTok and Instagram accounts and gave up the fight (Figure 8.6).

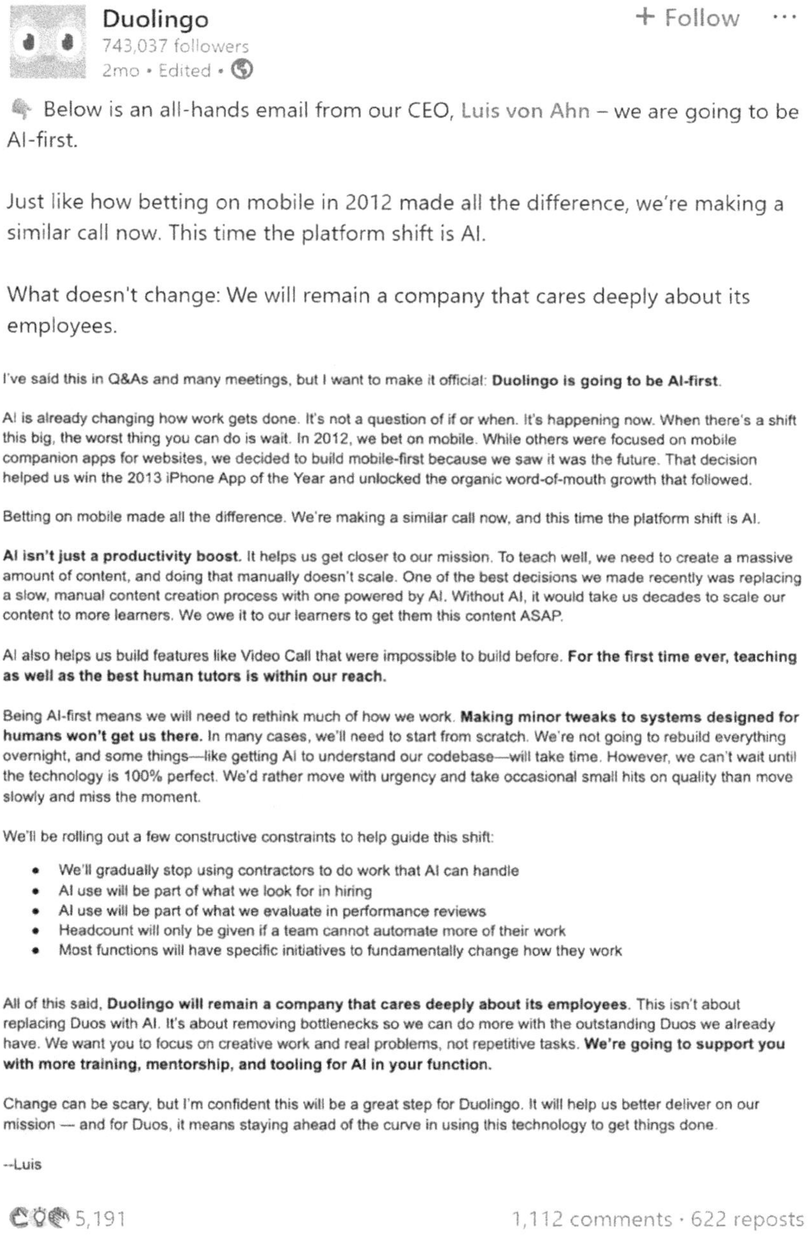

Duolingo
743,037 followers
2mo • Edited • 🌐

+ Follow   ⋯

👇 Below is an all-hands email from our CEO, Luis von Ahn – we are going to be AI-first.

Just like how betting on mobile in 2012 made all the difference, we're making a similar call now. This time the platform shift is AI.

What doesn't change: We will remain a company that cares deeply about its employees.

I've said this in Q&As and many meetings, but I want to make it official: **Duolingo is going to be AI-first.**

AI is already changing how work gets done. It's not a question of if or when. It's happening now. When there's a shift this big, the worst thing you can do is wait. In 2012, we bet on mobile. While others were focused on mobile companion apps for websites, we decided to build mobile-first because we saw it was the future. That decision helped us win the 2013 iPhone App of the Year and unlocked the organic word-of-mouth growth that followed.

Betting on mobile made all the difference. We're making a similar call now, and this time the platform shift is AI.

AI isn't just a productivity boost. It helps us get closer to our mission. To teach well, we need to create a massive amount of content, and doing that manually doesn't scale. One of the best decisions we made recently was replacing a slow, manual content creation process with one powered by AI. Without AI, it would take us decades to scale our content to more learners. We owe it to our learners to get them this content ASAP.

AI also helps us build features like Video Call that were impossible to build before. **For the first time ever, teaching as well as the best human tutors is within our reach.**

Being AI-first means we will need to rethink much of how we work. **Making minor tweaks to systems designed for humans won't get us there.** In many cases, we'll need to start from scratch. We're not going to rebuild everything overnight, and some things—like getting AI to understand our codebase—will take time. However, we can't wait until the technology is 100% perfect. We'd rather move with urgency and take occasional small hits on quality than move slowly and miss the moment.

We'll be rolling out a few constructive constraints to help guide this shift:

- We'll gradually stop using contractors to do work that AI can handle
- AI use will be part of what we look for in hiring
- AI use will be part of what we evaluate in performance reviews
- Headcount will only be given if a team cannot automate more of their work
- Most functions will have specific initiatives to fundamentally change how they work

All of this said, **Duolingo will remain a company that cares deeply about its employees.** This isn't about replacing Duos with AI. It's about removing bottlenecks so we can do more with the outstanding Duos we already have. We want you to focus on creative work and real problems, not repetitive tasks. **We're going to support you with more training, mentorship, and tooling for AI in your function.**

Change can be scary, but I'm confident this will be a great step for Duolingo. It will help us better deliver on our mission — and for Duos, it means staying ahead of the curve in using this technology to get things done.

--Luis

👍🩵❤️ 5,191   1,112 comments · 622 reposts

FIGURE 8.4
Duolingo email on being AI-First.

Luis von Ahn [in] · 3rd+   + Follow · · ·
CEO and co-Founder at Duolingo
2mo · 🌐

Today marks the largest expansion of courses in Duolingo's history: we've
launched 148 new language courses—more than doubling our offering
overnight.

These new courses are not new languages. They are languages we currently
teach that are now available to users of other base languages (for example, X for
speakers of Y).

This is a milestone I'm proud of. Our first 100 courses took us about 12 years to
develop. These 148 new courses? Less than one year. 😅

This acceleration is possible because we've combined our learning expertise with
advances in generative AI. We've created a way to build a high-quality base
course and quickly customize it for dozens of different languages.

We're closer than ever to making quality education universally available. The
technology that once seemed like it would take decades to build is here today.

https://lnkd.in/ehJKtESj

**Duolingo Launches 148 New Language Courses |
Duolingo, Inc.**
investors.duolingo.com

❤️👍💬 3,179   142 comments · 66 reposts

FIGURE 8.5
Duolingo's new 148 courses using AI.

Eventually Duolingo's CEO published a statement appearing to backtrack on
the AI decision and saying tech corporations need to work hand in hand with
AI to progress forward, as well as saying their mission statement remains the
same (Figure 8.7). Furthermore, he emphasised that they had no intention of
laying off employees in exchange for AI. This statement was met with a mixed
response from users and still attracted significant criticism. Overall, this was a
poor attempt to mitigate the AI affair, as the company elected to still utilise AI,
stated that they would use it for the betterment of the corporation and viewed
it as an essential part of the future. The ideal response would have been to do
a full reversal or to focus on the ethics and morality of the technology rather
than their future profits, as their statement implies.

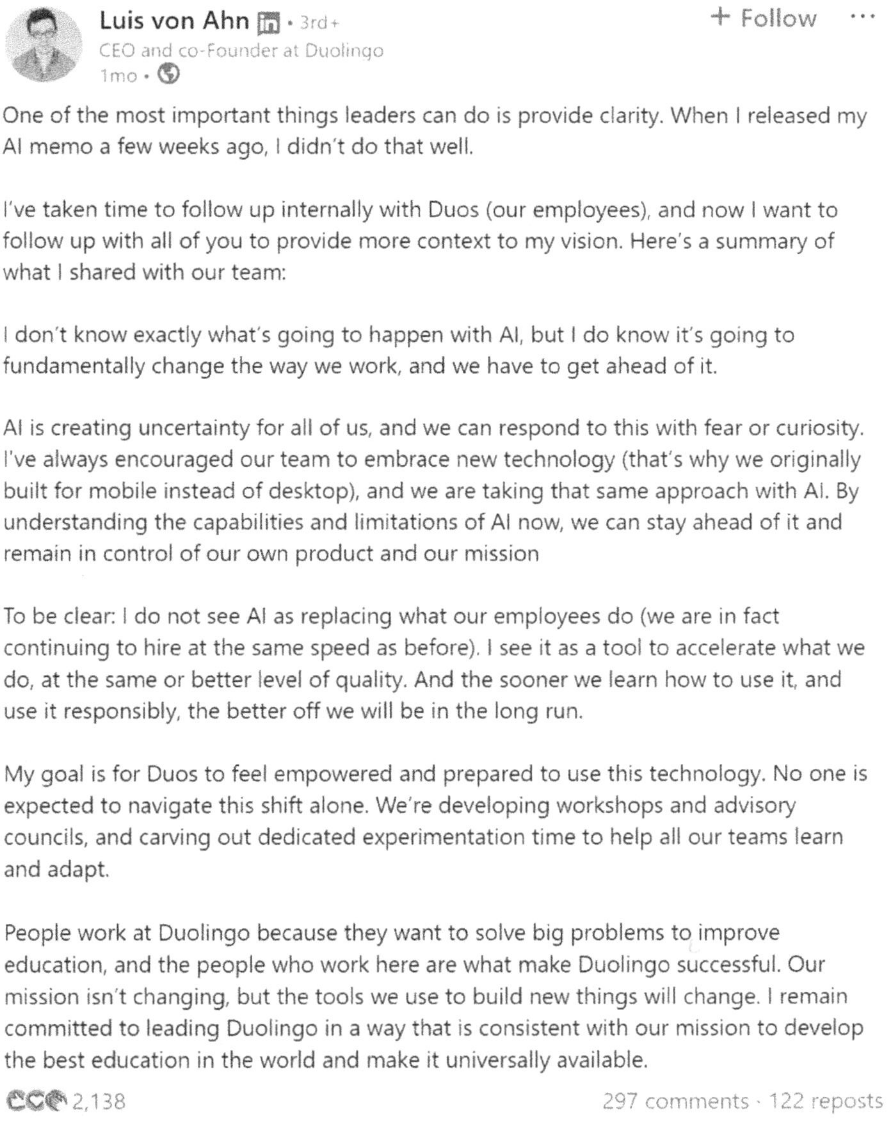

Luis von Ahn [in] · 3rd+
CEO and co-Founder at Duolingo
1mo · 🌐

+ Follow ···

One of the most important things leaders can do is provide clarity. When I released my AI memo a few weeks ago, I didn't do that well.

I've taken time to follow up internally with Duos (our employees), and now I want to follow up with all of you to provide more context to my vision. Here's a summary of what I shared with our team:

I don't know exactly what's going to happen with AI, but I do know it's going to fundamentally change the way we work, and we have to get ahead of it.

AI is creating uncertainty for all of us, and we can respond to this with fear or curiosity. I've always encouraged our team to embrace new technology (that's why we originally built for mobile instead of desktop), and we are taking that same approach with AI. By understanding the capabilities and limitations of AI now, we can stay ahead of it and remain in control of our own product and our mission

To be clear: I do not see AI as replacing what our employees do (we are in fact continuing to hire at the same speed as before). I see it as a tool to accelerate what we do, at the same or better level of quality. And the sooner we learn how to use it, and use it responsibly, the better off we will be in the long run.

My goal is for Duos to feel empowered and prepared to use this technology. No one is expected to navigate this shift alone. We're developing workshops and advisory councils, and carving out dedicated experimentation time to help all our teams learn and adapt.

People work at Duolingo because they want to solve big problems to improve education, and the people who work here are what make Duolingo successful. Our mission isn't changing, but the tools we use to build new things will change. I remain committed to leading Duolingo in a way that is consistent with our mission to develop the best education in the world and make it universally available.

👍👏❤ 2,138

297 comments · 122 reposts

FIGURE 8.6
Duolingo CEO clarifying the usage of AI.

The good – the archive strikes back

The final issues to be covered are the discovery of OpenAI stealing numerous works from the largest fanfiction website currently in existence, Archive of Our Own, also known as AO3. This site is ad-free, open source and funded entirely by its vast userbase. AO3 has attracted significant criticism for its

With the proliferation of AI tools in recent months, many fans have voiced concerns regarding data scraping and AI-generated works, and how these developments can affect AO3. We share your concerns. We'd like to share what we've been doing to combat data scraping and what our current policies on the subject of AI are.

Data scraping and AO3 fanworks

We've put in place certain **technical measures to hinder large-scale data scraping** on AO3, such as rate limiting, and we're constantly monitoring our traffic for signs of abusive data collection. We do not make exceptions for researchers or those wishing to create datasets. However, we don't have a policy against responsible data collection — such as those done by academic researchers, fans backing up works to Wayback Machine or Google's search indexing. **Putting systems in place that attempt to block *all* scraping would be difficult or impossible without also blocking legitimate uses of the site.**

With that said, it is an unfortunate reality that anything that is publicly available online can be used for reasons other than its initial intended purposes. In many cases, AI data collection traffic relies on the same techniques as the legitimate use cases above.

Once we became aware that data from AO3 was being included in the Common Crawl dataset — which is used to train AI such as ChatGPT — we put code in place in December 2022 requesting Common Crawl not scrape the Archive again.

We cannot go back in time to stop data collection that already occurred, or remove AO3's content from existing datasets, as much as we may dislike that it happened. All we can do is attempt to reduce such collection in the future. **The Archive's development team will continue to be on the lookout for individual scrapers collecting AO3 data, and to take action as needed.**

Likewise, our Legal committee has and will continue to serve the OTW mission of protecting fanworks from legal challenge and commercial exploitation. This includes their position that users should be allowed to opt out from having their works incorporated into AI training sets, a position that they have presented to the U.S. Copyright Office. They, too, will continue to keep pace with this developing field.

What can I do to avoid data scraping?

You may want to restrict your work to Archive users only. While this will not block every potential scraper, it should provide some protection against large-scale scraping.

AI-generated works and AO3 policies

At the moment, there is nothing in our Terms of Service that prohibits fanworks that are fully or partly generated with AI tools from being posted to the AO3, if they otherwise qualify as fanworks.

Our goals as an organization include maximum inclusivity of fanworks. This means not only the best fanworks, or the most popular fanworks, but all the fanworks that we can preserve. If fans are using AI to generate fanworks, then our current position is that this is also a type of work that is within our mandate to preserve.

Depending on the circumstances, AI-generated works could violate our anti-spam policies (e.g. if a creator posts a significant number in a short time). If you're uncertain whether a work violates our Terms of Service, you may always report it to our Policy & Abuse team using the link at the bottom of any page, and they can investigate.

This statement reflects AO3's policy at the time of writing, as we wanted to be transparent with our users about what our current stance is and what can be done - and is being done - to mitigate scraping for AI datasets. However, these policies are also under discussion internally among AO3 volunteers. If we agree on changes to these in the future, those will be announced publicly; additionally, if there are any proposed changes to the AO3 Terms of Service, they will be made available for public comment as is required of any and all changes to our Terms of Service.

We hope that this helps to make things more clear - this is a complicated situation, and we're doing our very best to address it in a way that doesn't compromise AO3's principles of maximum fanwork inclusivity or legitimate uses of the site. As discussions and approaches evolve, we will keep our users updated.

FIGURE 8.7
AO3 Statement on AI.

lack of censorship. However, it was discovered that the entire site had been data scraped for its extensive amount of creative works so that these could be fed into AI learning models.

This scraping was brought to light by Reddit user Kafetheresu who discovered this by using a tool called Sudowrite, a writing app that utilises GPT-3. Kafetheresu used Sudowrite to generate specific content that is very prevalent on AO3, these topics ranging from Omegaverse fanfiction, a niche genre focused around Alpha and Omega dynamics, to fanfiction involving the popular K-Pop band BTS. Upon this discovery, Kafetheresu immediately contacted AO3 to share their findings.

AO3 immediately took steps to ensure their users' creative works were given significant protection to prevent future data scraping and restrict works from being fed into AI generators. This is a measured and correct response, which also stems from the fact that AO3 is community-based and community-run. They recognise that without their users, they would cease to exist; a fact companies need to remember: no customers, no company.

However, AO3 does allow AI-generated works on its site to preserve artwork and under the acknowledgement that whilst AI-generated, it is still art. Any attempt to restrict its posting would conflict with their policy of having little censorship. However, AO3 does not require users to label these works as AI-generated. But as AO3 has an extremely extensive tagging system, often avoiding these works is rather trivial, as users tag the work regardless. However, given the controversy around AI across the artist and writer community, AO3 took a flexible stance whilst the law is decided.

Furthermore, when a dataset containing AO3 stories was posted on the dataset site Hugging Face in March 2025 (Figure 8.7), AO3 community members promptly contacted the administrators of Hugging Face and had the dataset taken down (Nyuuzyou, 2025). So, whilst the problem does persist, AO3 has shown that protecting their users and their works is an important priority for them and that such protections from AI and data scraping can be achieved.

What is the correct corporate response?

The discussion of these examples now prompts the following question: what is the correct response? One tactic currently in use is to neither publicise nor comment on the use of AI, which allows employees to use ChatGPT and Copilot at their discretion. Whilst this stealthier approach does shield a corporation to some extent, there is the problem that some employee rushing to meet a deadline may leave part of the AI output on their work, such as 'Sure, here is your revised press statement'. Corporations cannot hide their involvement in AI forever. Therefore, the correct response is not to replace workers with AI in the name of cost savings, as this is likely to lead to a backlash and consequent loss of revenue, but to use AI in a more ethical way. One such step would be to apply models to data sourced ethically and use the system in conjunction with humans. Another step would be to limit the use of systems like Copilot and ChatGPT in the workplace, or if employees insist on using them, then ensure that they are checked by an additional human to ensure there are no mistakes. Significant steps should be taken to avoid being complacent towards AI systems and their outputs. Furthermore, AI systems should not be used to replace significant aspects of the business, as in the case of Duolingo, or to introduce a system that completely negates the point of the business, such as in the case of NaNoWriMo, where users pointed out that the AI tools applied could finish a whole book in a day, making NaNoWriMo effectively pointless. Instead, AI needs to be only used as a way to speed up and increase the efficiency of the business, through using more ethical AIs trained on ethically sourced datasets. Essentially AI systems must be implemented in a way that does not damage the corporation or its employees as well as ensures the support of their customers, users and stakeholders. Overall, the corporate response to AI has been a curious lesson in the alienation of users, which leads into the next issue, the users themselves.

Kester Brewin offer his perspective on some of these issues.

EXPERT REFLECTION

Kester Brewin, *Associate Director at the Institute for the Future of Work, UK*

AI AND ITS IMPACT ON SOCIETY

We live in a socio-technical world and always have done. The tools we co-create perform technical functions that loop back and alter the ways in which we gather and interact. Consider a hammer: it amplifies my reach in the world, allowing me to do something that I cannot do with my bare hands: putting nails into a wall. But the hammer also reaches back through the arm into the heart and asks me a question: what do I want to *do* with this increased power? This is a socio-political question.

The history of industrial progress is thus a history of how new technologies have demanded reorganisations of our life together. Resonating with Polanyi's concept of the 'double movement' – whereby the expansion of a market is met by a self-protective countermovement of society to check that expansion – innovation has been matched by the formation of collectives of workers to steer that innovation towards good. With early hints in the Roman *Collegia*, it is in the guilds of the Middle Ages that we see social structures form to share, formalise and assure skills and techniques in working with new innovations. At their best, they supported the growth of economies by aiding diffusion of better practice across a membership body; at their worst, they were closed shops that sought to protect exclusive access to markets and suppressed further innovation as a means of protecting a certain social group. However, it is in the Industrial Revolution that we see these socio-technical impacts amplifying. With more powerful machines, more profound political, social and moral questions are raised, and existing social structures are put under greater strain as the powers of this new tool are assimilated. Trades unions were a societal response to technological innovation: how could we gather together to put the rights of people above the rights of the machine owners… whilst also remaining positive about innovations that make work less dull, dangerous and dirty?

And so to AI.

Here is a set of tools promising an exponential rise in the level of amplification: we will be able to do so much more. But with this greater power comes significantly greater pressure on the ways we relate to one another. We are already seeing major shifts in economic relationships: What is creative work worth? What reach should algorithmic decision-making systems have? What resources is it appropriate to sacrifice to build machines that might relieve us of toil?

The questions AI is presenting us are more urgent, more political and have greater ethical implications – but the structures that we have to debate our corporate responses have been hollowed out. In Bowling Alone (2000), Robert D. Putnam examines the gradual erosion of the meso-level institutions that would have facilitated this. We no longer gather in the same way in churches, in trade unions, in bowling leagues, or in political groups. This marks the present moment as one of particular significance: we have created one of the most extraordinarily powerful tools in the history of humankind, one that is likely to precipitate the most profound changes in our socio-economic relationships. But our ability to act together and reflect on how we should reconfigure our society to be able to thrive alongside this great innovation has been gravely weakened. Our corporate conversation is almost entirely dominated by powerful corporations, emperors in chinos telling us how it will be and how great life will be again if we yield to their more powerful intelligence.

AI's impact on society ought not to be a conclusion foregone by a small number of innovators. In the interplay between tool creation and social formation, there has always been this double movement as technologies are iterated and communities then question, respond and reform. What is incumbent on us now is to refuse to accept having our future narrated to us. We must gather and reflect and insist on our agency. This is what society has always done and what we must find ways to do more urgently than ever before: as AI impacts society, society must counterpunch and have an impact on AI.

The negligence of the user

The negligence of the user is regarded as the user not knowing or realising the consequences of the technology they are using. This negligence is not the fault of the user entirely, due to companies attaching long and confusing terms and conditions to their products. Furthermore, some companies choose not to publish changelogs, a log of exactly what is changed within their software, leading to further confusion. For example, if we examine almost every up-to-date videogame published in modern times, with every update, gaming companies release a comprehensive list of all changes made and all bug fixes, even if it is as simple as fixing a crash or adding a new character to the game. This is what is known as a changelog. It is presented to the users in a very obvious way. In contrast, social media companies either do not release changelogs or hide them from the user beneath a vast network of submenus.

The importance of this to AI is that users are unaware of the harm the technology does. This harm ranges from damaging artists and taking over jobs to the climate impact of this technology. The way to mitigate this is for companies to provide more transparency as well as effective education about the impact of AI. For example, most people can generate an image just for fun but have no idea that this action takes the energy of a fully charged smartphone (Tangermann, 2023). So the question to be asked is, where does this end? The main problem is that AI is taking over some of the easier jobs within corporations, posts that trainees or college leavers, apprentices and university graduates traditionally filled. Traditionally trainees would gain valuable experience and skills before moving up the organisation. The increasing use of AI means fewer trainees and students entering the industries for which they have studied since there is much greater competition against both their peers and systems like ChatGPT and Copilot. For example, there is the issue of when a senior staff member moves on and there is no viable replacement. The natural endpoint of this is a shortage of senior staff and the need for students, trainees and junior staff to become more experienced so they can fill more responsible senior positions. Another perspective is that the junior staff moving to senior roles are still reliant on AI input so that their expertise and knowledge are compromised. Therefore, these AIs may have a long-term negative impact on these companies.

Furthermore, there is also the negative impact that AI is having on the education sector and on the university lifestyle. Students are increasingly utilising AI to save time on assignments or to fully complete their assignments. Whilst saving time using systems like Grammarly or ChatGPT to check English or change reference formats from Harvard to APA 7 is a suitable use of AI for assignments, using AI to write the essay is blatant cheating. Yet this has become widespread within universities. The effect of this is that students become increasingly reliant on systems like Copilot and ChatGPT, thereby compromising their levels of learning, understanding and capability.

The problem is that when these students are faced with a situation or a problem where they cannot use these systems, it means that they may not be able to cope or fix the problem. Therefore, it then calls into question the value of a degree if it has indeed mainly been completed by AI. This type of degree has already been named as a GPT-degree or a GPT-powered degree. This flagrant disregard for academic standards and rules needs to be addressed. Universities face a choice to either amend their rules to adopt a more flexible AI stance or update their rules to be more restrictive on AI use and adapt their courses and assessments to ensure that AI is of minimal help. This could be achieved through an increased focus on practicality and real-world experience such as extracurricular activities, team building and an increased focus on industry and placement programmes.

There is also significant ignorance about the energy consumption of AI systems. For example, it took the equivalent of the annual energy consumption of 160 American houses to train ChaptGPT-4 (Aibin, 2024). Therefore, it is no surprise that Google and other tech companies are funding nuclear power plants in order to further their goals for AI. This means that if Big Tech gets priority for energy, then consumers will be affected during times of high demand. This becomes even more of a problem when a data centre is kept online at the expense of basic necessities such as water pumps or energy supplies to houses. The negligence of the user will increasingly cause such issues due to a lack of knowledge about resources that data centres use. Therefore, to mitigate this, there must be better regulation of the power grid to ensure that necessities are prioritised and that Big Tech does not control the energy reserves.

Overall, the negligence and ignorance of the users is problematic, as it will allow Big Tech corporations to have almost unchecked powers and influence regarding AI. Furthermore, as the users are unaware of the damage that they do, they will be less mindful of the way that AI is used. During the discovery of high levels of CO_2 emissions heating the atmosphere, BP launched a campaign blaming it on the general public, with other companies carrying out similar blame games. This stands in stark contrast to the response regarding Freon. This chemical compound was developed by Thomas Midgley Jr. as a safer alternative to ammonia, a dangerous and highly combustible chemical used in refrigerators. The use of Freon also expanded to aerosols and a range of other products. Yet it was soon discovered to be burning holes in the ozone layer. Governments and the scientific community acted quickly to impose a ban on Freon and began to phase out its usage over a ten-year period. This succeeded in preventing any more damage and allowed the ozone layer to gradually recover. Therefore, users need to be given a similar challenge to ensure sufficient AI regulation so that AI does not cause a net negative impact on the world due to poor governance. However, the danger is that with greenhouse gas emissions, there is a point of no return, and whilst AI may be used to help mitigate a variety of problems, there is the issue that it contributes significantly to global warming.

AI poisoning

In an effort to eliminate the data scraping, an organisation known as TheGlazeProject has developed tools such as Glaze and Nightshade to change artworks to ensure that they are not copied by AI art generators. Glaze works by adding minimalist changes to the artwork, essentially like an invisible filter which an AI will interpret as something completely different, for example, changing a realistic charcoal drawing into a modern abstract piece. Therefore, when the image is fed into the AI during the training process, it learns from a completely different image, protecting the artist's work. This is in contrast to Nightshade, which actively poisons the AI by using similar techniques yet turns the image into an incomprehensible mess so that when the AI is trained, its effectiveness is decreased. This can be so effective that it can lead AI image generators to return broken images in the form of grey screens with ripples and other poor outputs. The main difference between these systems is that Glaze is purely a defensive system that transforms one image to look like another, whilst Nightshade is offensive and actively damages the AI's performance. Generally, both of these tools should be used together for the best protection.

Furthermore, there are masks such as Chameleon which are designed for more personal photos such as selfies. The way Chameleon works is very similar to Glaze, in that it uses an invisible filter to embed certain features into the image that facial recognition models will recognise and classify as someone else (Chow et al., 2024). Yet, this does potentially mean that there will be misclassification if the image actually resembles another real-life person.

When considering face recognition technology, there are similar mitigation techniques here. Notably, the manifesto collection, a line of clothing developed by cap_able (capable.design, 2025). These jumpers, sweaters, and other clothing are designed to fool facial recognition. Furthermore, there are always facemasks which can conceal your face fully. There have been incidents involving making a mask with specific facial features and images on the front which fool facial recognition into misclassifying one person as another. Yet the issue of the same person being classified over and over does pose an issue. If that person is potentially identified by cameras as being at a crime scene and no one realises that the person is fake or misclassified, then there arises a problem. This can lead to a miscarriage of justice if law enforcement arrests the wrong person based on misclassification or time wasting if law enforcement spends time searching for a fake person.

However, if multi-spectral cameras with high resolution are used, then the effectiveness of masks is somewhat limited, as the cameras are able to penetrate them, although the usage of multi-spectrum cameras is currently very limited and extremely costly. A more interesting approach to AI poisoning is the use of .ass files. .ass is a little-known filetype used for subtitles like .srt which is the default filetype used for subtitles. The way .ass is different is it allows extensive formatting options. These formatting options allow users to place invisible text beneath the subtitles such as an article on ducks, for example. YouTube currently has a wave of AI-generated summary videos infecting its platform, and sometimes, they are generated from videos by other creators. Therefore, if YouTube creators use .ass subtitling on their videos, then the summarising LLMs are confused by the .ass subtitles; therefore, they will generate a video about ducks or whatever subject is hidden in the subtitles rather than the actual video content (f4mi, 2025). If this is implemented widely, then it is likely to deter most people from using generated content once the poison is discovered and therefore likely to reduce the amount of AI-generated content on YouTube.

Whilst AI poisoning is a good solution for combatting data scraping and AI companies stealing artwork, there is the future issue of AI eventually being able to bypass these systems or, indeed, detect poisoned data in the dataset and remove it. Furthermore, the main issue with data poisoning is that whilst it is very useful for artists, when other data poisoning techniques like labelling elements of the dataset incorrectly and deleting crucial elements of the dataset are applied to datasets such as those for medical or engineering data, there is a significant negative impact on the models. AI poisoning when applied to critically important datasets can lead models to make incorrect conclusions. This can be particularly damaging for medical datasets, since it will increase the probability of making incorrect diagnoses and therefore limit them significantly. Whilst this can be corrected with data processing as well as other mitigations, it is still a pressing concern for future AI modelling, particularly if the AI poisoning is carried out aggressively. In these cases hackers often add poor data to the dataset or sabotage the dataset in some way, rather than simply poisoning the data that is gathered to make the dataset. Yet this hacking of datasets also leads to potential other forms of AI poisoning. For example, hackers could steal a dataset and replace it with a poisoned one whilst holding the clean dataset for ransom as a form of a ransomware attack. Overall, AI poisoning is a great option for protecting artwork and other creative works, but there are potential ethical and security concerns when it is applied to critical datasets.

The agony of opt out

Whilst AI poisoning can be used for protecting your data, the main reason for the popularity of AI poisoning tools like Nightshade and Glaze is because of the inability to opt out of having data being used for training AI datasets. One such example is Meta, who planned to data scrape all of their services such as Facebook and Instagram for use in their Llama Large Language Models. Users who lived in the European Union were provided with an opt-out clause due to EU law, yet those in the USA were given no such choice (Heikkilä, 2024).

This becomes a further issue when considering the number of terms and conditions that everyone signs when downloading applications and signing up for accounts. One similar example of this is when Microsoft quietly rolled out AI systems for Word Documents and Excel Notebooks and, according to Microsoft, left it enabled on default by accident (Shilov, 2024). To disable this feature required trailing through a series of submenus to find a small, unremarkable tick box that reads 'Enable optional connected experience'.

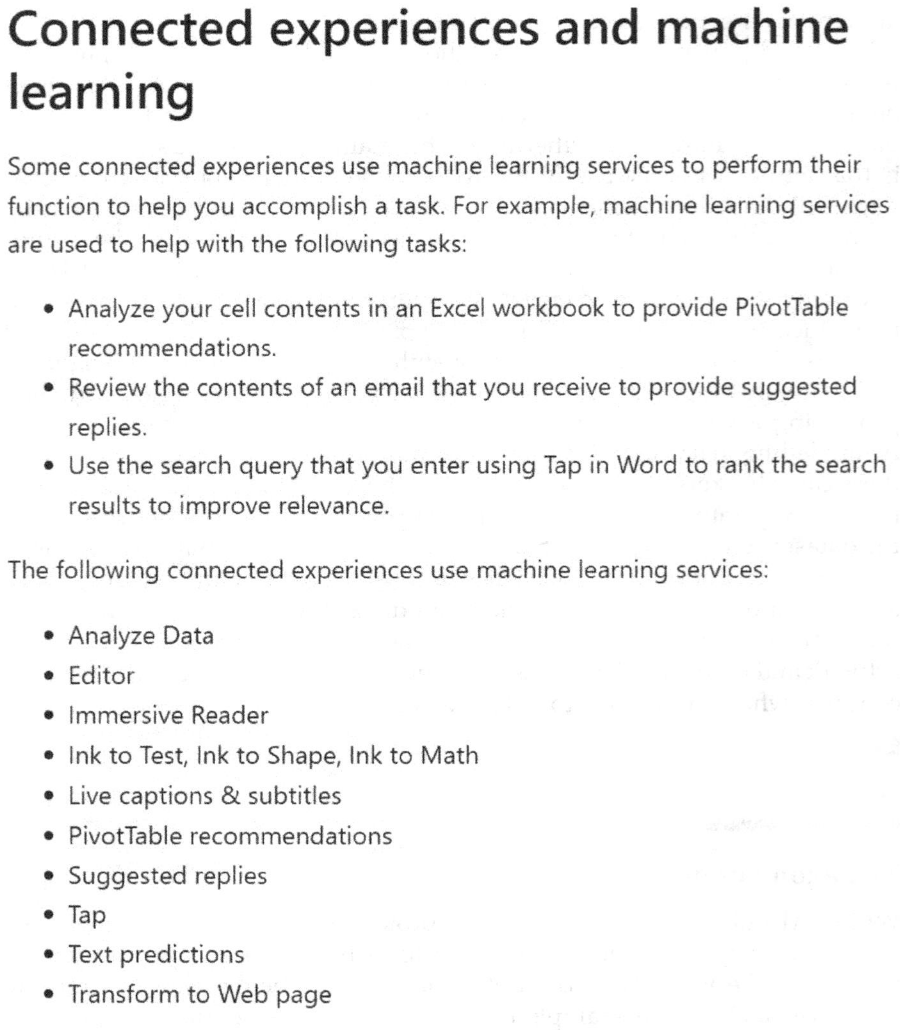

Connected experiences and machine learning

Some connected experiences use machine learning services to perform their function to help you accomplish a task. For example, machine learning services are used to help with the following tasks:

- Analyze your cell contents in an Excel workbook to provide PivotTable recommendations.
- Review the contents of an email that you receive to provide suggested replies.
- Use the search query that you enter using Tap in Word to rank the search results to improve relevance.

The following connected experiences use machine learning services:

- Analyze Data
- Editor
- Immersive Reader
- Ink to Test, Ink to Shape, Ink to Math
- Live captions & subtitles
- PivotTable recommendations
- Suggested replies
- Tap
- Text predictions
- Transform to Web page

FIGURE 8.8
Connected experiences option.

This wording is vague to say the least and the published Microsoft page on connected experiences, although it shows exactly which tools utilise Machine Learning, is also vague on how the data is handled and processed (Microsoft, 2024b). Furthermore, the page also stresses that Microsoft employees cannot access the data but makes no comment on whether AI algorithms can or not (see Figures 8.8 and 8.9). Yet they later refuted the claim that this system uses AI (Shilov, 2024).

How we use personal data

Microsoft uses the data we collect to provide you with rich, interactive experiences. In particular, we use data to:

- Provide our products, which includes updating, securing, and troubleshooting, as well as providing support. It also includes sharing data, when it is required to provide the service or carry out the transactions you request.
- Improve and develop our products.
- Personalise our products and make recommendations.
- Advertise and market to you, which includes sending promotional communications, targeting advertising, and presenting you with relevant offers.

We also use the data to operate our business, which includes analysing our performance, meeting our legal obligations, developing our workforce and doing research.

In carrying out these purposes, we combine data we collect from different contexts (for example, from your use of two Microsoft products) or obtain from third parties to give you a more seamless, consistent and personalised experience, to make informed business decisions, and for other legitimate purposes.

Our processing of personal data for these purposes includes both automated and manual (human) methods of processing. Our automated methods often are related to and supported by our manual methods. For example, to build, train, and improve the accuracy of our automated methods of processing (including artificial intelligence or AI), we manually review some of the output produced by the automated methods against the underlying data.

As part of our efforts to improve and develop our products, we may use your data to develop and train our AI models. Learn more here.

FIGURE 8.9
Microsoft privacy statement extract.

However, all doubt is dispelled when examining Microsoft's privacy statement (see Figure 8.10), where it clearly states that it can use customer personal data for AI training (Microsoft, 2024a). Whether it obtains the data from Word is unclear, but considering the amount of text written in Microsoft Word, it seems only a matter of time. A further interesting aspect of this privacy statement is that upon clicking the 'Learn more here', redirects to Microsoft Copilot. This is questionable, as Microsoft then entrusts a large AI

> **11. GOVERNING LAW.**
>
> For Services provided to, on, or in orbit around the planet Earth or the Moon, this Agreement and any disputes between us arising out of or related to this Agreement, including disputes regarding arbitrability ("Disputes") will be governed by and construed in accordance with the laws of the State of Texas in the United States. For Services provided on Mars, or in transit to Mars via Starship or other spacecraft, the parties recognize Mars as a free planet and that no Earth-based government has authority or sovereignty over Martian activities. Accordingly, Disputes will be settled through self-governing principles, established in good faith, at the time of Martian settlement.

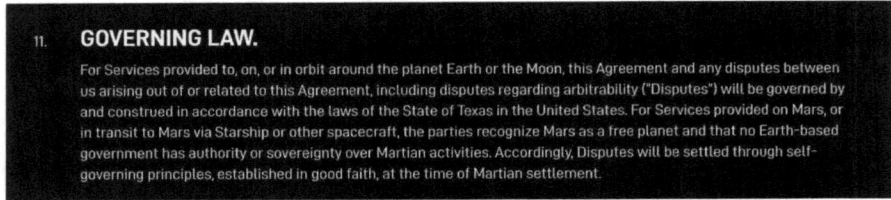

FIGURE 8.10
Governing Law section of Starlink's terms of service.

language model to tell customers about how their data is used for AI training. Furthermore, it is interesting to consider that whilst disabling connected experiences does disable Word and Excel's AI features, does it also really disable Word's data scraping?

Overall, Microsoft and other Big Tech corporations need to provide more transparency as to whether they utilise their users' data for AI training. Legislation needs to become more comprehensive so as to ensure that companies provide sufficient notification to their users and in more obvious places than buried within submenus. In the case of Grammarly, a writing tool used to correct grammar, it is a simple yes/no slider in settings, unlike with Meta and Microsoft, where the terms are buried behind a labyrinth of submenus with confusing names. The difficulty with terms of services being hidden is that they can be abused and/or used for political gain. One such example is the terms of service of Starlink which contains the following section, as seen in Figure 8.10 (Starlink, 2024).

This section essentially stipulates that any activities on Mars should be settled through a Martian settlement, and no earth government has jurisdiction over Starlink systems deployed on Mars. This can be read as an affirmation of Martian independence at the time when a Martian settlement is established, and, whilst it can be read as a joke, it is still legally binding and means that users signed up to Starlink have to recognise Mars as a free planet. Yet, this does contravene the Outer Space Treaty (UNOOSA, 1966), which recognises celestial bodies as being under international law, meaning that this particular section can be classed as void (Brown, 2020; Salmeri, 2020). However, the true impact of this is unlikely to be known until a Martian settlement is established. The point is, if tech corporations can introduce sections like this into their terms of service with minimal legal repercussions, then it is a growing concern of what can be included in terms of service without users being aware of them. Therefore, the transparency and contents of these terms must be put under further scrutiny and regulation.

Whilst sparse regulation of terms of service continues, there are already tools such as Terms of Service Didn't Read (TosDR), which provides summaries of the service terms and grades based on how fair the terms are, and Claudette (Lippi et al., 2019), which uses AI to identify unfair terms of service. However, the issue here is that the user is required to be proactive in checking the terms of service using these tools. Yet the paper by Obar and Oeldorf-Hirsch (2020) found that the majority of users, even those users concerned with privacy spend less than a minute reading the terms of service. The time most users spend on terms of service is scrolling to the bottom to click accept. Therefore, this points to a genuine need for these summaries as well as ranking tools to be implemented into the first pop-up inviting users to accept the terms of service. Furthermore, these rankings should also be accompanied by the company's stance on AI as well as other information specific to the AI system the corporation utilises. Nevertheless, the existence of tools like TosDR is certainly a good step in the right direction.

Conclusion

The main issue addressed in this chapter is that users of AI systems are unaware and, to some extent, negligent of the effects that AI has on the world. Therefore, the main solution to this is resources that serve to educate users and the general populace to ensure that they are aware of these consequences. Furthermore, when corporations and governments are concerned, users need to be holding them to account, as well as taking steps to show that AI use is not acceptable to them, such as boycotting their services, as in the case of NaNoWriMo. This should help push corporations towards the correct response, like AO3. It is interesting that a potential future tactic for corporations could be to just not use AI in order to gain customers. There needs to be an overall push towards the increased regulation of Big Tech in the fields of AI and data protection, when it comes to both terms of service and the way that the AI systems are used. Overall, whilst a great push for AI regulation is needed, when this does not go far enough, there are tools that allow the users their own measure of regulation. The current realistic options for the users to take measures into their own hands are by utilising AI poisoning techniques.

9

The panic of the future and the need for dissent

Introduction

This chapter begins by exploring the way in which future predictions are resulting in new forms of panic. It begins by examining just what kind of panics we are dealing with and then compares this with predictions for AI and human expertise. The second section of this chapter explores forms of dissent; these forms of dissent illustrate the impact of AI on academic discourse and political, social and digital dissent. The final section explores dissenting acts and the ways in which AI helps and hinders dissent in different spaces. The argument of this chapter is that there needs to be a stronger sense of, and action towards, dissent in AI in this digital age.

AI versus human expertise

Innovation and the advancement of technology mean that the general public and the less educated are often left behind. This has occurred throughout history such as during the Industrial Revolution and since the invention of the computer. Some of these advancements have been accompanied by government skills programmes, for example in the UK Diplomas in Accounting and additional Advanced level (A-level) qualifications have been offered by the National Careers Service. Yet, none of the previous technological advancements have had anything like the scale and effect of AI, although it could be argued that there are similarities with the impact of the assembly line in terms of efficiency and reducing production time. The first assembly line was introduced by Henry Ford, which cut the Ford Model T car production time down from twelve hours to a mere hour and a half (Encyclopedia Britannica, 2024). Despite the introduction of increased wages and fewer work hours for the employees, a significant number were laid off and forced to find work

DOI: 10.1201/9781003622949-10

elsewhere. Some unionised in an attempt to fight for better support for the workers, whilst most simply found jobs elsewhere, often outside the automotive industry.

The recent rise in Generative AI feels more than just an option for improving skills since LLMs such as ChatGPT and Copilot can write entire reports, press releases and even write code in mere minutes. Such automation has resulted in a significant efficacy in the amount of work completed but has led to workers being made redundant. This surge in the use of AI may not last since AI makes mistakes that result in the need for proofreaders, fact checkers and programmers to check its code. However, very specific and specialised codebases are either impossible to use by AI or are specifically restricted from being fed into LLMs. Therefore, a certain amount of skills development of humans will be required. Furthermore, the perceived need for the mass integration of AI across various industries will be likely to cause a brief boom in the requirement for AI specialists and data scientists to implement and maintain these systems, especially given that LLMs can often make mistakes or miss certain syntax errors. This becomes especially prevalent when LLMs are provided with a very long programme to fix or a long document to summarise, due to their limited memory potentially leading to later context errors. Therefore, workers can be helped to develop their capabilities to assist the AI with a project too large for its memory or to proofread the programme to correct errors and bugs. However, the growing problem with AI is that equipping more capable workers with new skills will result in fewer jobs being accessible for less skilled workers. Whilst at one level this can be used to improve people's skills and capabilities, Germain and Grenier (2022) also suggest that it can result in laziness and a lack of creativity. For example, if tasks can be completed by AI, workers equipped with AI will ask it to do the task every time, and AI will not necessarily provide the best or most creative approach to the problem.

An interesting insight is the lack of trust shown in AI identified by Agarwal et al. (2023) regarding the analysis of chest X-rays. This paper explores how performance varied between humans, AI-assisted humans and solely the AI system. The researchers found that whilst AI was better at diagnosis than two-thirds of the radiologists, providing AI insight to the radiologists did not lead to the radiologists performing better. This was due to the radiologists not trusting the AI system and therefore not valuing its opinion. This led to most radiologists taking more time to come to a decision. The solution, the paper surmised, was that AI should work alongside humans rather than with them. Thus, radiologists should not be provided with the AI tool initially but later in the diagnostic process, in more of a colleague format rather than simply as a verification system. If the conclusion of this paper is extended to the real world, then it is probable that worker expertise will remain the same unless the AI system codes are edited to a more collaborative form or more trust is placed in AI. At this point there is likely to be less job reduction, due to the majority of the workforce still being needed as AI is treated as a colleague rather than a replacement. Yet it is interesting how much skills development will be facilitated by the AI and as the technology improves, how more AI colleagues will spread throughout industries. A survey published by Grace et al. (2018) explored the timeframe of High-Level Machine Intelligence (HMLI). A High-Level Machine Intelligence as defined by the paper is 'when unaided machines can accomplish every task better and more cheaply than human workers' (p. 3). The conclusions drawn indicated that 55% of the people surveyed believed that AI would have a good outcome overall. Furthermore, the survey concluded that there was nearly a 75% probability of HMLI being developed in the year 2116, one hundred years after the survey was conducted in 2016. It is at this point that the main concern will be for employment. However, one important consideration is the idea of human employment being cheaper, as AI models require significant time, water and energy to generate data. Furthermore, in the idea of autonomous worker robots as depicted in the TV show and the game Humans, Detroit: Become Human (Brackley & Vincent, 2015–2018; Quantic Dream, 2018) and many other media, there will always be a need for humans to ensure that these robots are serviced and maintained as illustrated in Figure 9.1, below.

Figure 9.1 shows the predictions of when AI will be able to achieve certain tasks based on the survey. A significant number of these milestones such as Go, High School Essays and Translation/Transcribing have already been achieved a few years early. Yet, these achievements have failed to assist the extension of human expertise. In the case of Go, it was the primary reason behind top player Lee Sedol quitting the sport after he felt disheartened by his loss against OpenAI's AlphaGo (Pranam, 2019). In the case of education, it is a hindrance to students who submit AI-written essays since their learning is limited by doing this. Yet, it can help them with generating ideas if they so choose, allowing them to develop searching capabilities. However, the increase in the usage of AI-generated essays by dishonest students places unnecessary stress and pressure on honest students to perform better, which may increase their expertise yet at the expense of their mental health. This stress may lead the student's opinion to slowly turn to 'If you can't beat em, join em' and therefore destroying their commitment to the idea of increasing their expertise as they instead choose to submit work produced by AI. When it comes to language, AI has a little further to go due to mistranslation and errors with certain dialects but will enable some increase in human expertise as translators begin to adopt more varied and niche languages that are unfamiliar to an AI. Ironically, with the mass adoption of niche and varied languages by translators, this means that a surge of more accurate translation data will be available and therefore will make it significantly easier for AI to learn these languages, making it easier for it to then take over this translation. Furthermore, AI has also seen massive strides in sign language and can allow a greater level of accessibility due to better translation software.

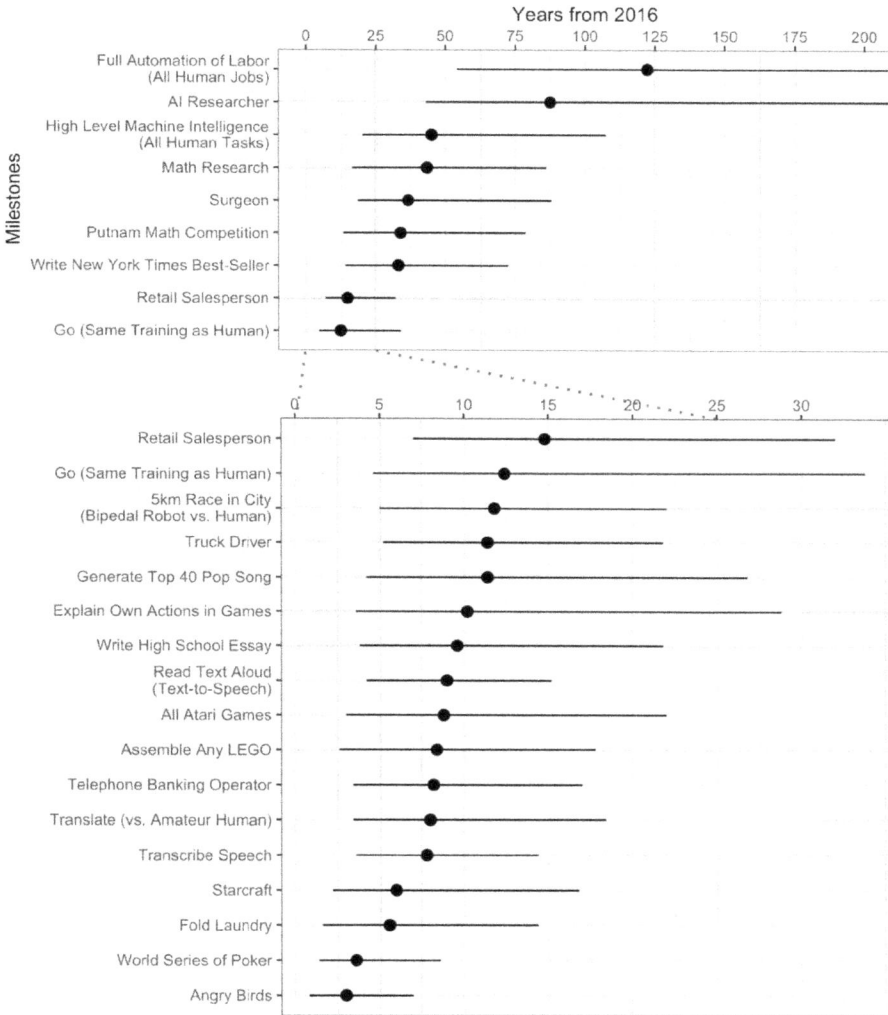

FIGURE 9.1
Timeline of median estimates (with 50% intervals) for AI achieving human performance (Grace et al., 2018) (Reproduced with permission).

Yet as sign language is a rather personal language of communication, it is more than likely that the human touch will always be preferred by those using it.

In conclusion, AI can encourage an increase in human expertise and capability within certain industries. In other industries where the AI excels, there is a likelihood that the need for human expertise will be eliminated; therefore, its ability to increase in expertise is limited only to the industries it cannot yet automate.

Forms of panic

With the idea of AI being so advanced that it can almost replace the human worker, as previously discussed, there comes a certain panic due to the potential rise of poverty despite the economic prosperity that AI could bring.

Loss of jobs

The idealistic view of this future is something akin to Star Trek (Roddenberry, 1966) or The Orville (MacFarlane, 2017) where everything is provided and hence there is no need for money. The more realistic perspective might resemble the television series The Expanse (Corey et al., 2015) where most of the population is on a welfare programme that provides basic housing, poor healthcare and poor-quality food. Yet, neither of these fictional systems came to fruition without the significant joblessness, economic hardship, wars and infighting that came before. With mass automation of jobs also comes the risk of economic disaster should these systems suffer a mass failure or become hacked in some way. Therefore, the idea of a reserve workforce to be utilised in case of disaster would be a good mitigation.

Implant technology

Another form of panic comes from the advancement in implant technology and the idea of brain emulation. Though experimental and often implanted by tattoo artists or body piercers, various technological implants are in use around the world. These body modifications or 'body mods' enable the users to augment their abilities, such as being able to sense the electromagnetic spectrum, to turn their office lights on and off with a thought. These body mods can be as simple as an RFID chip implanted in the hand to allow contactless payment and better device integration

The company Neuralink has conducted a variety of experiments with brain implants on monkeys, which, despite significant setbacks and deaths, successfully allowed the monkeys to play the virtual version of ping pong known as pong. When Neuralink's chip was approved for human trials in May 2023, the company wasted no time in implanting one inside Noland Arbaugh, who was paralysed from the neck down. This chip allows him to mentally control a computer, granting him some small freedoms, allowing him to play games and search the web freely (Kleeman, 2025). A year later, the chip is still operational, and Nolan reports no side effects. Yet, Neuralink is far from the only company pushing ahead with this technology; researchers from Lausanne University in Switzerland successfully placed brain implants into Gert-Jan Oksam, who was paralysed from the waist down in a cycling accident. These implants allow him to control the implant in his spine, allowing him to walk (Ghosh, 2018). The Chinese Institute for Brain Research and the NeuCyber NeuroTech have also seen significant success with their three patients able to control robotic arms with enough dexterity to pour cups of water without spilling them and the ability to display their thoughts on a screen. There is also the possibility of uploading memories into the chip, which would be beneficial for the justice system and witness statements as well as providing a potential mitigation to the effects of dementia.

Misuse of technology

Whilst this technology is overwhelming and impressive, technology could be misused and perhaps offer a rather dystopian worldview. What happens if these chips get hacked? Will there be significant brain damage if the chip fails? What privacy concerns will there be? Will adware be streamed directly into the brain? In the digital age, there are mass privacy concerns everywhere, especially with the idea of smart devices that can connect to everything. Whilst the idea of controlling your house with your mind seems appealing, getting hacked by your smart fridge is not, and whilst smart devices only see data, the brain implant sees your thoughts, memories and emotions. There is also the concern of how this most precious data is stored and secured wherever possible to prevent abuse. Neuralink has claimed that they cannot see the personal data from the chip implanted in Noland Arbaugh, yet despite this, he says that he is taking some care on what he thinks through the chip (Kleeman, 2025).

Furthermore, if all these brain chips are made compatible with each other, then a more horrifying thought comes to mind: the idea of telepathy. Although this could be quite handy for greeting friends each morning, this also leads to the idea of accidentally oversharing and all sorts of interpersonal problems. Furthermore, an enemy could just broadcast insults into your head constantly, and there could also be a significant level of perversion such as digital catcalling. Therefore, significant safety and security measures must be implemented such as firewalls, being able to block other chips' access and, most importantly, being able to turn the host chip off entirely. Furthermore, with the nature of AI data gathering, which often involves data scraping without permission, will corporations resort to scraping memories and thoughts from these chips in order to fuel the next generation of AI models?

Forms of dissent

The idea of dissent is based on the assumption that there is a need for change. Those who seek such change, who question, or who express diverse or alter- native views are catalysts for new ideas and new ways of thinking. This may appear to be subversion, something that is looked on unfavourably, but in fact this is to misunderstand what it means to subvert. Whilst the word *subvert* is typically understood as undermining the authority and power of an institution or system or ideology by overthrowing and destroying, subvert is actually derived from the Latin *subvertere*, from *sub-* *'from below'* + *vertere* *'to turn'*. So, to subvert is to turn from below. In this way, we want to subvert commonplace practices and ways of being and codes of conduct and expectations for success by turning them upside down like turning over a rock in order to see what is typically not seen, to reimagine other possibilities for the rock. The current state of play is that there is little turning over or reimagining occurring about what can be done to interrupt the illegalities associated with AI. However, there are forms of dissent that can help us to begin the work of resisting the negative and unlawful practices seen in many forms of AI and these are presented below.

Academic dissent

Academia has, possibly until recent years, always been a place that sought to challenge the status quo. In the 2020s this is probably less so as universities increasingly come under government scrutiny. Davids (2021) notes that disagreement and dissent should be at the heart of the university ethos, indicating that conflicting truths have a right to be heard. She explains:

> The university ought to be the one place where the expression of views and positions – no matter how controversial – ought to be heard. Disagreement and dissent are critical to the purpose and responsibility of a university. The moment at which academics, students or speakers are cut off, prevented from speaking or disinvited, or the moment academics are instructed to use this, and not that text, the university becomes closed off to new forms of engagement. The more universities shut down ideas – under guises of fear of causing offence or for issues of safety – the more insular and uncritical the academic and institutional spaces become. The point of engaging with controversial arguments is not simply to listen to that which might be harmful or despicable, but to create opportunities for talking-back – that is, to engage in reasoning, and to pursue truth.

(Davids, 2021, p. 1188)

Universities worldwide are focussed on performance, and in order to improve this, they have appointed professional managers who are intolerant of dissent. Such managers are dedicated to growth, financial performance and improving university rankings, despite evidence that this does not improve university performance or student engagement. For example, some years ago, Pelz and Andrews' study of 1,300 scientists found that those who were most effective were those who valued their freedom, pursued their own ideas, did both basic and applied research and were also able to influence decision makers (Pelz & Andrews, 1966). Furthermore, Jain et al. (2010) building on this work, argue that dissent is vital for creativity in research teams. Dissent tolerance in universities in the US has also been examined through modelling and simulation (Zaini et al., 2019). The findings indicated that silent staff, authoritarian administration and university management resulted in the decline in the university's performance. In terms of AI, there is little dissent in terms of the use of AI-driven plagiarism software, such as Turnitin, with few academics standing against this, but there is a realisation by academics that students' submission of AI-generated assignments is beginning to destroy the ability in students to think and critique. Helping students to understand cheating and fake news now needs to be included in university teaching, across all subjects, so that universities are able to encourage the development of critical beings.

Political dissent

Political dissent requires that people speak out about the extent to which governments restrict connectivity and block social media platforms. It is evident that particularly during elections, governments block websites that host political and social speech. It is also vital that people require their governments to uphold transparency to ensure that humans review decisions made by AI algorithms such as asylum decisions and health and social care provisions.

Social dissent

From a social perspective, people need to be aware of, and stand against, the currently allowed surveillance mechanisms. Examples of such surveillance include spyware, face recognition software and predictive policing, which can result in repression and the invasion of human rights. The difficulty is that such practices remain covert and are therefore difficult to challenge. What is needed are more effective laws on digital privacy. In practice, people should demand that governments pass laws on purpose limitation so that personal data gathered for one purpose cannot be used for another, as this will reduce the reuse of harvested information that occurs without people's consent.

Digital dissent

We define digital dissent as an expectation, and to some degree a requirement, that people will defend a free and transparent Internet based on principles of democracy. In practice, this means protecting human rights, preventing misuse of data and arguing for the release of people imprisoned or targeted because of online expression against a country. We suggest that digital dissent will also require that laws be repealed that unfairly criminalise online expression. To date there is little discussion or even definition of digital dissent, but in terms of the realistic and ethical use of AI, this needs to be at the forefront of discussions in this area.

Radical dissent

The notion of radical dissent is firmly located in the Baptist faith. Christianity developed out of Judaism when Jesus came to challenge norms and offer a different view. In the Roman Empire, all religions were accepted as long as they proclaimed the emperor to be a god. Christians refused to do this, and such subversion was seen as an act of radical dissent. Radical dissent emerged from the Baptist movement as they became critical of the church and society, arguing for the need for religious liberty rather than an imposed religious ideology. Radical dissent came in the form of resistance to state-imposed religious conformity in the 19th century. In the 21st century, radical dissent means the rejection of state dictates and any form of totalitarianism. Freedom to choose, opposition to discrimination, ecological responsibility and rejection of the abundance of possessions are central tenets of radical dissent. The implications for this in terms of AI are that religious organisations need to stand up for religious freedom worldwide and prompt the state to consider ethical and ecological ways of using AI.

Dissent in the face of AI

The pervasive nature of AI and the worldwide differences between government stances on it mean that dissent is deeply problematic. Perhaps a useful concept here is that of heterotopias: the idea that spaces are not only other-worldly but worlds within worlds; spaces are at once mirrors of reality and spaces of interruption. Foucault's delineation of heterotopias ranges from prisons and asylums to brothels and baths (Foucault & Miskowiec, 1986) – which perhaps reflects the state of capitalist global businesses where dissent is very much frowned upon. Heterotopias are seen as unsettling spaces because they illuminate the extraordinary in the ordinary. Foucault's heterotopias, whilst wide-ranging in concept, are nonetheless best understood as ones that seek to locate and invent cultural and social spaces (Foucault, 2008). The implication for AI is the need to consider the way in which AI might be used as a positive means of dissent and an 'other' space, as well as recognising that AI can be used to undermine a just society.

Sarah Hayes offers her perspective on AI below

EXPERT REFLECTION

Sarah Hayes, *Professor of Education & Research Lead in the School of Education, Bath Spa University, UK*

PRESERVING POSITIONALITY, RELATIONALITY, CONSENT AND DISSENT

In order to bring a certain clarity to the array of challenges presented by artificial intelligence (AI) and to avoid the silos of excessive optimism or negativity, it is first necessary to resist 'reducing AI to a singular, hyped entity' (Raffaghelli et al., 2025 p. 20). AI primarily concerns our *data* (in all of its biodigital forms) and our *positionality* as individuals who make choices in our personal, postdigital contexts (Hayes, 2021). However, our choices are always *relational* in a world of 'co-constitution' (Gergen, 2009; Perrotta & Selwyn, 2019). Therefore, secondly, even as we explore the relational possibilities of the many forms that AI takes, it is vital for our wellbeing and that of the planet to protect our own 'generative processes of relating' (Gergen, 2009). What, for example, are the creative responses and human interactions that we seek to preserve and develop as we adapt and relate to generative artificial intelligence (GAI)? A consideration of what may be lost or preserved brings personal, positional responsibilities towards ourselves, each other and the wider community.

In this respect, a helpful framework to refer to is that of Human Data Interaction (HDI). This conceptual framework (Mortier et al., 2014) consists of four core tenets. Firstly, *legibility* suggests that data and analytics algorithms ought to be both transparent and comprehensible to the people who have data and processing concerns. Secondly, *agency* argues for people to have the capacity to act within these data systems, to opt-in or opt-out and to control, inform and correct such data and its inferences. Thirdly, *negotiability* relates to changing relationships around data and data processing, including our individual attitudes and how society responds to these through different legal or regulatory frameworks. Finally, *resistance*, or dissent, needs to be preserved as a basic human right (Hayes et al., 2023, Turvey, 2024).

In summary then, the question of how each of us seeks to protect our own 'generative processes of relating' (Gergen, 2009) to the AIs we brush up against, via the key tenets of HDI, is both an individual/positional and a collective response. It is necessary to 'question the epistemologies that shape our coexistence as we encounter and strive to enrich our postdigital positionalities as academics through a conversation that embraces both technological and human diversity' (Raffaghelli et al., 2025, p. 22). It is likely, then, that for some time yet, both fear and hope will oscillate for each of us, as we seek to protect our relational positionalities, to consent or dissent, amid our many AI-Human Data Interactions.

Dissenting acts

Whilst in general in this chapter we see dissent in the face of AI as a positive step, there are many dissenting acts in a digital age that have unhelpful and negative consequences. These include grandstanding, gaslighting and scapegoating, but we cover just three others here: whistleblowing, public shaming and the digital pillory.

Whistleblowing – in the main whistle blowing has proved to be an effective means of challenging harmful or illegal practices. Whistleblowing occurs when someone who works inside an organisation obtains information through their own access, hacking documentary analysis or hidden video recording. Examples of these are the Snowden and Luxembourg leaks and the Cambridge Analytica scandal.

Edward Snowden was working for the National Security Agency and chose to disclose its surveillance to a documentary filmmaker in 2013. His findings revealed that government agencies in the USA were able to ask companies for copies of their data (Snowden, 2014). The Luxembourg leaks were initiated by whistleblower Antoine Deltour, who showed evidence of widespread tax avoidance in 2014. Despite the development of new policies by the EU, it is clear that when business is conducted virtually, taxation becomes a complex area. As Olesen (2022) notes, tax havens and offshoring strategies, whilst illegal, do exploit any grey areas that exist. In the Cambridge Analytica scandal, it was claimed the company used Facebook data on behalf of its clients in order to send political messages to their users who could then be influenced. On March 17, 2018, the *Guardian* and the *New York Times* broke the story, based on information received from a Cambridge Analytica whistleblower, stating that the company had accessed 50 million profiles of Facebook users for their modelling. Facebook referred to this as a data breach but in fact the data was not actually stolen since it was gained with users' consent. Zuckerberg, the founder of Facebook introduced restrictions on developers' access to data as well as a tool for all users to be able to see which apps could access their data. The impact of this scandal has been that some social media have banned microtargeting and advertising. This controversy has since had a huge impact on data privacy, political campaigning and social media. Hu (2020) argues that there is a need to explore the potential for protection within the enforcement actions by regulatory agencies. Governments worldwide have enacted laws and regulations to protect consumers, but we suggest that these still do not go far enough ethically.

Public shaming – in the past, public shaming included putting people in the stocks, as well as flogging, branding and cutting off the hand of a thief. Public shaming is the act of public criticism when someone is deemed to have transgressed social norms. Today, public shaming includes placing offenders' names in newspapers and social media and posting signs on people's doors or fences. People who engage in public shaming draw attention to some kind of violation, often posting it online, in order to not only draw attention to it but also to rally others to their cause. This is more commonly referred to as 'Cancel Culture'. However, there are more extreme examples, such as a mother in Ohio who chose to punish her daughter for speaking disrespectfully to her in front of her friends. She placed an X over her daughter's Facebook profile picture with a message next to the picture saying, 'I do not know how to keep [my mouth] shut. I am no longer allowed on Facebook or my phone. Please ask why'. The mother made her daughter answer every inquiry in the belief that this sort of punishment would have some sort of positive effect on her daughter (Goldman, 2015, p. 439). The difficulty with AI in this is the tendency to reinforce norms that are not always justified, along with an inability to forgive past actions. Furthermore, as Billingham and Parr (2019, p. 26) note, online shaming is also often 'questionably accurate'.

The digital pillory – is the practice of shaming ordinary people when they commit some kind of minor offence or mistake, such as drunkenness or urinating in a public place. It is a practice that normally occurs when someone is filmed in the offending act and then this goes viral. It also falls under the umbrella of Cancel Culture. Hess and Waller (2014) explain this through the example of Kerry Ann Strasser urinating on her seat at the end of a football game in Brisbane. Captured on a phone camera and uploaded to YouTube, this went viral in hours. More recently, in the UK, Mary Bale was labelled the 'cat bin lady' as she dumped her cat in a wheely bin. Just over half (54%) of the population believe Mary Bale, dubbed 'cat bin lady', has received the right punishment for her actions, being fined £250. She also was publicly shamed and lost her job. Mobile phones are clearly creating a new form of public surveillance and the ability to pillory people though this kind of surveillance is reinforcing the assumption that treating miscreants in this way is acceptable. It seems that our societies are becoming increasingly unforgiving despite apologies being made and even before legal action has been taken.

In order to create a just society, there needs to be a deeper consideration of one's own moral stance rather than blaming or cancelling other people.

The importance of criticality and critique

Being critical, speaking out and having a voice in 21st-century society can often result in condemnation and being 'cancelled'. The idea of being cancelled is that someone is ostracised for a particular stance or view. The difficulty is that Cancel Culture can prevent free speech and the opportunity for open debate, and, as Ng points out, 'digital practices often follow a trajectory of being initially embraced as empowering to being denounced as emblematic of digital ills' (Ng, 2020, p. 621).

In a digital world where critical thinking and critique are important, it is vital that people are able to explore their stances and views and not be silenced. Barnett (1994) has argued for three levels of criticality, which have been adapted here as follows:

Critical thinking – this is the process of someone thinking through an issue carefully and critically to decide on their stance towards an issue.

Critical thought – Critical thought is collaborative and has a wider focus than just the individual's thought processes, seeking to explore collective perspectives. Critical thought necessarily contains a social component and thus can only be developed collaboratively. For example, a debate around the dinner table on the practices of the far right will enable people to consider their own stance whilst debating with others.

Critique – In critique, different views of an issue or situation may be proffered as alternative perspectives are taken on board. This is a real cognitive and personal challenge, and it may open up the way to a transformation in someone's perspective. In many ways this can be seen as a hardline debate where people seek to explore, research and understand issues in-depth, thereby developing a well-rounded understanding and an ability to see beyond their own biases. Table 9.1 considers these stances in relation to AI.

However, despite needing to understand and take a stance towards AI, it is important that critique should not be used as a weapon. Fitzpatrick (2021) suggests that there is a contagious version of critical thinking that negates creativity and is characterised by a refusal to listen, which in turn means it becomes corrosive. The spread of neoliberal values across the globe means that critical thinking can be weaponised to promote the wants of the individual over the collective good. What is needed is the valuing of the sort of critical thinking, critical thought and critique that is open, creative and inclusive and not the kind that is used as a weapon and which devalues humanity and reason in a just society.

TABLE 9.1

Progressive criticality for AI

Level of criticality	Features	Stance towards AI
Critical thinking	Development of autonomy Use reasoning, analysis and synthesis	Seeking to understand one's own biases and stances towards AI
Critical thought	Collective debate and action Critical dialogue	Understanding AI in-depth in order to be able to debate the challenges that are created through its various uses
Critique	Criticism of the area under debate Taking a new stance towards knowledge and issues	Evaluating one's own biases towards AI through debate and examining personal stances to AI by deconstructing them and seeking new or different views

Conclusion

This chapter has argued that dissent is vital to the development of the realistic and ethical use of AI. Having considered the importance of valuing human expertise and examining new forms of panic, it has explored the ways in which the overuse and misuse of AI can be challenged through different forms of dissent. It is also important to consider that AI and social media in 2026 tend to reinforce often unhelpful and right-wing norms that need to be challenged through both dissent and a critical stance. Such reinforcement by right-wing advocates may be a passing fad, and in years to come, there may be other mechanisms that reinforce new forms of extremism and modernity.

10

Death, ethics and artificial intelligence

Introduction

Death remains a taboo subject across the globe, and yet, since the early 2000s, there has been a rise in advances and use of death tech. More recently, particularly following the COVID-19 pandemic, death and the digital have become increasingly intertwined. This chapter begins by exploring the death practices and the impact of AI, including the linking of AI to the brain. This chapter then moves on to examine the relationship between transhumanism and AI and then presents some of the current digital afterlife options. The final sections of this chapter explore death ethics and AI and some of the ethical conundrums related to this, including death prediction and AI, death and media waste and digital engagement with the dead trends.

Managing death

The recent trilogy by Shusterman begins with the novel Scythe (Shusterman, 2016). Here, we are presented with a world managed by a sentient artificially intelligent being, the Thunderhead, who watches and controls the world but does not control death. The world has become overpopulated because death is always curable, except by fire, so in order to manage the population, human scythes are a self-appointed group who 'glean' humans in order to manage the population. The allegedly sentient Thunderhead, whilst seeming not to interfere, does manipulate humans to a degree to try to manage the more evil scythes. Whilst this story might seem worlds away from the 21st century, humanity can and does manage death but in discrete and hidden ways through assisted dying, voluntary euthanasia and voluntary suicide. In recent years in the UK there has been much discussion about legalising euthanasia, with assisted dying being defined as prescribing life-ending drugs for terminally ill, mentally competent adults to administer themselves after meeting

DOI: 10.1201/9781003622949-11

strict legal safeguards. Voluntary euthanasia describes a doctor directly administering life-ending drugs to a patient who has given consent, which is legal in some European countries but not the UK. Voluntary euthanasia is permitted in Belgium, Luxembourg and the Netherlands. Those who stand against the legalisation of these processes choose to use a more sinister term, that of voluntary suicide.

The idea of linking death with AI might at first seem somewhat tangential. However, the rise of griefbots, apps to support grief and the creation of digital immortals has resulted in interest in this area that has possibly surpassed the cryogenics movement. Cryogenics, or more accurately cryogenic cooling, can be traced back to the ancient Egyptian Empire around the 16th century BCE to 11th century BCE. Cryogenics became a scientific discipline in the 19th century when scientists were able to liquefy gases at extremely low temperatures. The origins of absolute zero were discovered in 1848 by Lord Kelvin (Jain et al., 2021). Cryogenics, thus, revolves around the production and maintenance of tissues at below freezing point. Cryonics is the 'freezing' of the human body in order to seek to suspend life with a hope for future reanimation. Although freezing cells and embryos in medicine is now common, effective cryopreservation still remains impossible. However, despite the unlikelihood of reanimation, many people still choose to have their entire body or their head frozen after death costing as much as £80,000. Currently, there are three facilities, one in Russia and two in the USA (Lawford, 2021). The current difficulty is that whilst mind upload (also referred to as cloning) might be a possibility, the ability to upload this into a brain that has been cryopreserved is not achievable, despite suggestions by the companies concerned that it is possible to merge computers with the human brain. Furthermore, as Mullock and Romanis (2023) point out, the contractual arrangement that a person has with a cryonics organisation seems to be for preservation and storage and not for reanimation. Mullock and Romanis note:

> While the general rule is that the body is not property, we argue that cryonic preservation could transform human remains into property. While the family initially has a strong claim in deciding what happens to the body in terms of how the body is disposed of, once the cryonic organization takes possession and exercises skill in preserving the body, the issue of ownership becomes unclear.
>
> *(Mullock & Romanis, 2023, p. 3)*

There have been a few cases where family members have disagreed about someone being frozen and others fighting over the future of the cryogen, particularly when a dead body does not count as property, but this could clearly change if a cryonic becomes classed as property, which will no doubt result in further legal and ethical difficulties. What has become more marked in recent years are different ways of linking AI to the brain, which are described below in Table 10.1.

TABLE 10.1

Some of the current iterations of linking AI to the brain

Name	Description	Examples	Difficulties
Bitbrain https://www.bitbrain.com	This company combines neuroscience and artificial intelligence with hardware to develop brain-sensing devices and monitoring technologies	Control of robotic devices, motor rehabilitation, brain to vehicle interface	Given the high use of this in motor rehabilitation as yet it is difficult to tell who will or will not benefit from this technology.
Emotiv https://www.emotiv.com	This is a bioinformatics company seeking to understand the brain using electroencephalography (EEG)	There are a variety of areas from gaming to robotics, defence and security.	It is not really clear from the website who and how this technology is being used in practice. It seems to be designed for everyday use by individuals looking to understand and improve their own brains.
Kernel https://www.kernel.com/	This is technology designed to measure the brain	Measurement of depression and cognitive decline	This is being used to develop brain-based biomarkers. Some of their studies seem to suggest that experiments currently show low reliability
Neurable https://www.neurable.com	This uses headphones to collect brain data	Designed to collect data but also eliminate distraction for the user when they are working.	They seem to be headphones that collect data for the company. It is not entirely clear if there are any other purposes
Neuralink https://www.neuralink.com/	This is an implant that allows people to control devices with their thoughts	Most of the work appears to have been done on animals so far	There does not seem to be much indication as yet that this has been undertaken effectively with humans
Next Mind https://www.next-mind.com	This is designed to enable you to control things with your brain	It lets you select items in a virtual scene with your brain.	It seems to be a virtual reality device for moving things in-world but does not appear to offer much else.
MELTIN MMI https://www.meltin.jp/en/	This is developing cyborg technology focused on merging humans and machines through bio-signal processing	They develop algorithms to analyse bio-signals accurately and create robotic mechanisms inspired by biological systems.	The main focus appears to be on the creation of medical neurorehabilitation devices to improve movement, but it is not clear, as yet, how this is cyborg technology.

The growth of companies such as those above in many ways would seem to have an underlying transhumanist stance.

Transhumanism and AI

Transhumanism is a stance in which it is argued that technology can be used to transcend the limitations of the human condition. In 1995, Pearce argued that transhumanists were committed to super-longevity, superintelligence and super wellbeing – in order to live longer lives with improved cognitive abilities and satisfaction (Pearce, 1995). Pearce suggested that there was a need for humans to work towards the abolition of suffering in all sentient life through genetic engineering, nanotechnology, pharmacology and neuro-surgery. Roden (2014) suggested four forms of technology that could achieve this, namely nanotechnology, biotechnology, information technology and cognitive science. There is a sense that transhumanism has not been taken entirely seriously, possibly because of the early emphasis on cryogenics. More recently, the adaptation of the body to include chips to operate home smart devices, or the addition of extra body parts as art installations, has perhaps not helped the desire by those in this field to be taken seriously. The result is a sense of transhumanism being perceived to be an area of myths rather than philosophical and scientific ideas. For example, in the late 1990s and early 2000s, ideas about transhumanism related to some somewhat eccen-tric enhancements of humans. More recently authors such as Cohen and Spectator (2020) suggest an extreme version of transhumanism which seeks to dispense with the body by transferring, or uploading, the human mind from the biological brain to a computer. This, they suggest could facilitate travel within the solar system and enable galactic travel. However, Sigmund (2021) seems to offer a more realistic stance, suggesting that whilst transhu-manists may seek to enhance human abilities, this could be dangerous since humans are a balanced system and changes to some powers may destroy such a balance.

However, Rothblatt (2014) argued that it will eventually be possible for people to achieve a form of immortality through what is now termed the transhumanist 'church'. Based on transhumanism, the Terasem Movement Foundation was created. The Terasem Hypothesis, outlined below, is a means of people creating MindFiles through LifeNaut where anyone can create their own mindfile for free and includes the ability to create a photo-based avatar, which they believe will learn from conversations someone has with them. The Terasem Hypotheses state that:

1. A conscious analogue of a person may be created by combining suf-ficiently detailed data about the person (a 'mindfile') using future consciousness software ('mindware'), and

2. that such a conscious analogue can be downloaded into a biological or nanotechnological body to provide life experiences comparable to those of a typically birthed human.

Lifenaut enables people to create mind files by uploading pictures, videos and documents to a digital archive, but this is an explicit process. It also enables the user to create a photo-based avatar of the person that will speak for them, although there is the choice of only a single male and female voice, which are both US English, which, they argue, will allow them eventually to be reincarnated in a digital avatar. In practice, LifeNaut offers a number of ways to build a 'mindfile'. These include:

- Filling out interview questionnaires, including LifeNaut's Profile Personality assessments, which are many and varied; and in 2025 there are currently 21 new assessments.
- Talking to your own avatar so it learns from what you say, although this appears to require an explicit 'correction' action.
- Manually adding favourite things and URLs, although your bot does not appear to learn them once added.

It is however also possible to have your MindFiles beamed continually into space for later, potential, interception and re-creation by alien intelligences. The grandiosity of some of the claims made here are not all empirically testable, and it takes a huge amount of work to create a mindfile. In terms of AI, LifeNaut uses AI-powered chatbots to mimic a user's conversational style based on their uploaded content and AI-driven facial recognition to create lifelike avatars that can simulate expressions and speech. The hope in the future is that mind files can be uploaded to robots or even reanimated cryogens. The site states

> LifeNaut.com is a web based research project that allows anyone to create a digital back-up of their mind and genetic code. The ultimate goal of our research project is to explore the transfer of human consciousness to computers/robots and beyond.

> *(www.LifeNaut.com)*

The concern about the possibility of individuals surviving the death of their bodies or indeed being reanimated is, as Sparrow and Zhang (2025) note, that people will no longer fear their own death and will care even less about the death of others.

Digital afterlife options

As aforementioned, there are companies dedicated to creating digitally immortal personas and Facebook has now put in place measures to control the post-mortem data on their site (Brubaker & Callison-Burch, 2016). Steinhart (2014) has examined personality capture, mind uploading and levels of simulation, arguing for a computationally inspired theory of life after death. The idea of creating a digital afterlife is one that appears to raise issues and concerns for those in the field of software development. Maciel (2011) argues that seven issues reflect software developers' concerns about death: respectful legacy, funeral rites, the immaterial beyond death, death as an end, death as an adversity, death as an interdiction and the space required by death. In later work, Maciel and Pereira (2013) explored the beliefs around death within software development and found that religious and moral values affected sensitivities to the personification of death, which in turn affected design solutions. Studies such as this illustrate how cultural, political and religious beliefs can affect the technical landscapes around the design of digital afterlife.

Yet more recent interest has emerged in ways of leaving digital replacements of ourselves behind. These vary from griefbots to new forms of digital resurrection.

Griefbot

A griefbot is used to describe a bot that is created using a person's digital legacy from social media content, text messages and emails. The idea is to create a bot that replicates the deceased's interactions in order to help friends and family work through the grieving process (Villaronga, 2019). These chatbots are based on the digital footprint left behind by the deceased through social media, emails, texting and messaging systems – with the aim of providing the bereaved with the chance to speak to their loved ones after their death.

Dadbots

Journalist James Vlahos filmed his dying father, curated content about him and later created the *Dad-bot*, a chatbot of his dead father with whom he continues to communicate. The process of doing this prompted him to co-found HereAfter, a company and subsequent app that interviews you about your life, creates a virtual you and then enables loved ones to hear meaningful stories by chatting with you (Nast, 2017).

Digital afterlife and digital immortals

The concept of digital afterlife is defined here as the continuation of an active or passive digital presence after death (Savin-Baden et al., 2017). Other terms have been used to describe digital afterlife, including digital immortality. 'Afterlife' assumes a digital presence that may or may not continue to exist, whereas 'immortality' implies a presence, in some form at least, ad infinitum. Whilst these terms may be used interchangeably, afterlife is a broader and more flexible construct, as it does not contain assumptions about the duration or persistence of the digital presence (Savin-Baden & Mason-Robbie, 2020). Advances in data mining and artificial intelligence are now making an active presence after death possible, and the dead remain part of our lives as they live on in our digital devices (Bassett, 2015).

Digital immortals are where a virtual human is created that is to all intents and purposes indistinguishable from its physical progenitor. As such, and with the right hosting package, there is no reason why the virtual human, and the essence of any physical human it may be based on, should not become immortal. At its most basic, a digital immortal merely comprises code and data. Digital identities are data, which can be added to and updated (and even forgotten), and an application built from code with a set of rules (which may themselves be data) which enables the interaction between that data and the real world. The digital immortal can 'read' a variety of information sources and has two-way access to a range of real-world systems. As with any virtual human, it can potentially embody itself in virtual worlds (as an avatar) and possibly in the physical world (via a robot), as well as through a 2D interface or an email, social media, aural, chat or Skype presence. It has a natural language understanding and generation facility to enable two-way communication with people (and other virtual humans and digital immortals) in the physical world, and it potentially synchronises its knowledge and activities between multiple instances of itself.

Digital resurrection

Besser et al. (2023) argue that the desire for 'digital immortality is often initiated by the bereaved rather than by the deceased, regardless of their explicit consent' (p. 3). Examples of different forms of digital resurrection are provided in Table 10.2 below:

TABLE 10.2

Forms of digital resurrection

Company	Type of digital resurrection	Forms of AI being used	Reference
MyHeritage	Creation of moving GIFs based on photographs of dead ancestors	Deep Nostalgia, an AI-powered tool based on Deep Learning and computer vision using generative adversarial networks (GANs) and facial reenactment technology.	Wakefield (2021)
Shoah Foundation	Holocaust survivors testifying about their personal stories to allow conversations with their holograms after they die	Natural Language Processing and Machine Learning	Dimensions in Testimony (2020)
Amazon	Voice sampling to replace Alexa's voice with those of living or dying people	Neural voice cloning	Hern (2023)
MPL Communications	Extricating John Lennon's voice to add to unrecorded Beatles song	Custom-made AI by dialogue editor Emile de la Rey	Khomami (2023)

Whilst not technically a digital afterlife option, another iteration is the creation of digital companions for lonely people. Digital companions are virtual humans who are trained through Machine Learning to operate as companions. Such companions can be accessed on phones and computers and are designed to alleviate loneliness through conversations, online games and counselling. A number of small-scale studies have been undertaken in this area (Loveys et al., 2021; Ring et al, 2015; Ta et al., 2020) which suggest such companions may mitigate loneliness. However, some of the ethical concerns about the use of digital companions include addiction to the companion, dependency that reduces social interactions with other humans and concerns about data privacy of those using the companions (Jecker et al., 2024).

Death, ethics and AI

Questions about ethical memorialisation relate to whether digital legacies affect the way people feel about dying and the ways in which they choose to face or ignore death. For example, it is not clear whether the creation of digital immortals would mean that the deceased's funeral would be less meaningful or their memorial site less significant. Furthermore, it is still uncertain as to whether the digital immortals' effect will change bereavement practices, resulting in a different kind of continuing bond. Throughout most of the 20th century, it was expected that grieving involved a process of letting go, breaking bonds with the deceased. The continuing bonds model was developed by Klass et al. (1996) who argued instead that in fact the relationship with the deceased changes, rather than breaks. In short, the relationship does not end. It challenged the popular models of grief requiring the bereaved to detach from the deceased. Thus, the model suggests that grief is not a linear process to be worked through which is completed when someone 'moves on'. Instead, after death, the relationship is redefined, so that some kind of attachment is perceived to be normal. The focus is on a changed relationship, a continuing bond with the deceased, rather than a letting go and moving on. However, recently Klass has remarked that maintaining a connection with the dead is a common aspect of bereavement in all current models of grief, which is reflected in the use of digital and social media for this purpose (Klass, 2018).

A useful way of considering the impact of digital immortals is suggested by Buben (2015). Buben refers to Interactive Personality Constructs (IPCs) by which he means the capturing of someone's appearance, mannerisms, voice and memories, as well as their view of different topics. These are then synthesised through recording and motion-capture techniques. What Buben is referring to is still in the development stages through companies such as

Aura. Aura https://www.aura.page/ is a new site, created by Paul Jameson after he was diagnosed with motor neurone disease in 2017. Paul recorded his voice font before he became unable to speak and, in the process of dealing with this illness, began to work with family and colleagues to develop a site that would be useful for those facing death. Such developments are more likely to be referred to in 2026 as digital immortals rather than IPCs. However, Buben challenges us to consider when we are creating an IPC (or a post-death avatar) whether the purpose is for recollection or replacement, which I define, based on his ideas as

- *Recollection.* an awareness of the loss of the person and what has disappeared following their death, and the preservation of memories of the loved one.
- *Replacement.* the creation of a post-death avatar to replace the dead loved one.

The issues of replacement bring to the fore questions about whether creating a replacement results in people becoming less concerned about the death and loss of a loved one. Replacement would seem to be about filling a void or particular role and thus using the dead as a resource for the living. It might be the case that replacement results in distraction from loss and interrupts the process of grieving, since the focus on creating a replacement is to ensure that the 'dead stay around'. However, whatever is created, in whatever form, brings with it issues about the ethics associated with different kinds of memorialisation. Elaine Kasket offers her perspective on some of these issues.

EXPERT REFLECTION

Death, Ethics and AI

Elaine Kasket, *Visiting Professor at Centre for Death and Society at University of Bath*

We're constantly sold the idea that technology smooths inconvenience from our lives, soothes aversive emotions, and decreases our suffering. Might technology even allow us to defer or bypass death? Through commonly available Large Language Models (LLMs), many of us already possess the tools to create artificially intelligent chatbots of the dead.

The creeping normalisation of such AI applications is paralleled by the abnormalisation of loss. Popularising sophisticated, interactive representations of the dead risks reframing bereavement and grief as problems with technological solutions.

I would argue that being able to tolerate difficult experiences is fundamental to our psychological health and thriving, and I question the ethicality of technologies that divert us from struggling, suffering, and experiencing acceptance, growth, and resilience after someone dies.

But the ethical implications of AIs trained on digital remains goes beyond individual psychological considerations. 'High-fidelity' digital afterlives are disproportionately accessible to the privileged individuals already valued by society: the deceased CEO who continues to consult; the dead pop star on hologram tour; and today's thought leader who may stay on tomorrow's lecture circuit.

When we decide which digital personae to keep and which to cull, we maintain historical inequalities and reinforce existing power structures, allowing more marginalised voices to fade and degrade. When we choose to maintain the influence of select individuals through digital continuation and recreation, we also risk social, cultural, and ideological stagnation. 'Dead labour' could even reduce employment opportunities for the living.

The question of whose values dictate which digital remains to retain and amplify is bedevilling enough, but using those remains to train AI raises environmental concerns as well. To eke out one more year of time with an AI replica of a once-living person, we must drain a small swimming pool of the earth's water. Scaling up such practices is thumbing our noses at the climate crisis. These are high-carbon-footprint 'solutions' to a non-problem: all carbon-based life forms reach their end, making space and preserving resources for the generations to come.

When considering ethical dilemmas, we identify relevant stakeholders and how they might be affected by a course of action. The dead, one could argue, are not hurt by the manipulation of their digital remains, whether for emotional comfort, cultural capital, or financial gain. And yet, posthumous simulations and other forms of value extraction reduce the dead to tools, content, commodities. The more our digital reflections come to capture the essence of our personhood, the more reanimation practices should give us pause – especially in the usual situation of absence of consent.

Not all possible technologies are desirable, and grief and finitude are not glitches or user-experience 'pain points'. We should always inquire incisively about the problem for which we are trying to solve and be alert to unintended consequences. The mere fact that chatbots of the dead are infinitely more lifelike than they were only a short time ago illustrates that data rights, human rights, and dignitary rights should become far more intertwined than they currently are, in both life and death.

Ethical conundrums

Much of the debate that occurs in the public sphere about digital memori-alisation in society focuses on the legal concerns and the terms and condi-tions imposed by different platforms. However, the ethical concerns stretch beyond the legal concerns, since digital memorialisation relates to the social good and thus introduces questions about what counts as the social good and who decides? Can or should everyone have the right to a post-death ava-tar? What will occur if such avatars become sentient? Are there different lev-els of digital memorialisation that require different levels of ethical stances? Underpinning all of these questions is the need to delineate the values that inform the development, creation and management of digital memorialisa-tion, and the issue of values is central to this.

Kant (1870/1998) distinguishes between conditional values and uncondi-tional values. A conditional value is something valuable in some circum-stances, whereas an unconditional value is valuable in all. According to Kant, the value of, for example, intelligence is conditional because we can imagine circumstances in which it would be bad for someone to possess it, such as when it would be used for evil ends. By means of this line of argu-ment, Kant argues that 'the good will' is the only unconditional good, good in all circumstances. The good will is roughly a disposition to make morally commendable choices or to do the right thing.

The challenge then in terms of values is how artificial intelligence and digital immortals are likely to be shaped if there is no early intervention in the process of debate so that instead commercial and corporate values and interests have a free rein. If this is then pursued into the field of AI and digital immortals, there are some potentially disturbing implications about the power of the few to determine the values incorporated into these new technologies. In some ways, this could be said to be happening in the early 21st century already. Grandinetti (2020) notes funeral streaming services also collect considerable media from their viewers such as what is watched, where it is watched, how long it is watched and what kinds of devices are used. Streaming services then use data and connections for economic gain. Such economic advantage could be used to guide users towards films and resources about death, dying, bereavement resources and services before they have even buried their dead.

However, there are still difficulties about what is included in a digital immortal and what is omitted. For example, Newton remarked about the creation of a griefbot by a friend that:

> You may feel less comfortable with the idea of your texts serving as the basis for a bot in the afterlife - particularly if you are unable to review all the texts and social media posts beforehand. We present different

aspects of ourselves to different people, and after infusing a bot with all of your digital interactions, your loved ones may see sides of you that you never intended to reveal.

(Newton, 2016)

Reflections such as this illustrate the complexity of creating a digital immortal that may not be wanted by some of those left behind, as well as the difficulty of deciding who makes the choices about access, artefacts, indexing and ultimately when and how it is deleted.

Death prediction and AI

To date, in-depth discussion and research into death and AI are relatively uncommon. Many of the debates relate to the possibility and reliability of death prediction through AI, such as the Death Date Calculator Clock Timer. This is a free app that calculates when you will die and gives free life predictions and suggestions on the basis of your health, status, height, weight and lifestyle. The app not only calculates your death date but also suggests the solution to extend your remaining life and delay the death. More serious predictions are being used in healthcare such as a new algorithm developed by researchers at Stanford University (Wiederhold, 2019). The algorithm has the ability to predict the time of death within three to twelve months for a hospital inpatient. The process claims to be >90% accurate, and thus, hospital admission decisions in the future could be made on the basis of screening health records with artificial intelligence to determine the patient's need for palliative care before death. There are similar studies exploring the use of algorithms in areas such as heart disease and suicide prevention. For example, the healthcare provider Geisinger US used an AI system to examine 1.77 million electrocardiogram (ECG) results from nearly 400,000 people to predict who was at a higher risk of dying within the next year (Lu, 2019).

Death, AI and media waste

In 2019, Kasket argued that online persistence and the ongoing presence of the data of the dead online will lead to more of a globalised, secularised ancestor veneration culture, and it is important to recognise the ongoing persistence of the dead online on social media, LinkedIn, Amazon and YouTube. Yet huge media waste is created because people choose not to remove the absent. For example, Kasket suggests that Facebook may be hosting the digital remains of 4.9 million people by the end of the century. She argues:

In a 2019 UK YouGov survey, 16% of the 1,616 respondents said that they would like their profiles to remain online and visible to others after death, at least for a time (Ibbetson, 2019). A quarter of respondents, however, said that they would like all social media to be deleted entirely at the time of their deaths. This design is currently inexecutable, in both a practical and legal sense. Even if it were possible, would we seize the opportunity to make provisions for our digital estate when two-thirds of us already fail to make plans for our physical one (Chapman, 2018)?... Our personal data is simply too voluminous too widely spread throughout the datasphere and too under the control of innumerable third parties to be able to simply gather it up and 'bury' it even if that's what the deceased would have wanted.

(Kasket, 2021, p. 21)

In a study undertaken in 2019, before the onset of the COVID-19 pandemic, it was clear that many people did not want the digital remains of their dead friends and family deleted from social media sites (Savin-Baden, 2022). Yet, there are competing perspectives about the idea of retaining people in the digital. Data in 2024 indicated there are 193.8 million Facebook users in the United States, 378.05 million in India and that Facebook is expected to have 63.9 million deceased accounts by 2025 and 186.9 million by 2050. The growth on X (Twitter) has been slower, probably because of changes in ownership and organisational polices, but this is predicted to have 20.3 million deceased accounts in 2025, 61.1 million in 2050 and 94.9 million in 2100 (Chikwem, 2024).

Issues such as what to do with the dead in cyberspace suggest that absence and presence need to be viewed in a less binary way, as Derrida (1997) has suggested, absence resides within presence. Whilst many people consider those who have died to remain dead in cyberspace, for others, there is a sense of their continued presence as angels in a cyber dimension. Perceived absence then can be seen across a spectrum, from opting out of digital media through to relative absence in digital spaces to the valuing of the dead perceived to be alive in cyberspace – who are absent in life but have in fact left behind digital identities.

Death engagement trends

Morse (2024) argues that there are conflicting trends in engagement with death and the digital realm. The first is the use of everyday platforms that were not designed for grief and mourning as well as oppositional shunning of digital spaces especially designed for engaging with, or memorialisation of the dead. Morse, building on the work of Bassett (2022), suggests directions of communication between the living and the dead which are described and exemplified in Table 10.3 below:

TABLE 10.3

Directions of communication between the living and the dead (drawn from Morse, 2024)

Form of communication	Description	Example	Associated research
Planned one way communication: dead to the living	The dead reach out to the living through intentional left behind messages on a digital afterlife platform.	Before someone dies, they leave messages to be delivered posthumously on preselected occasions using sites such as MyWishes and SafeBeyond	Savin-Baden, (2022)
Accidental one way communication: dead to the living	This is accidental communication which is software generated	Algorithms in a social media site can accidentally promote interaction on a dead user's account	Bassett (2022)
One way communication: living to dead	The use of messaging services to communicate with the dead	Leaving messages on the deceased's Facebook timeline	Segerstad et al. (2022)
Intentional two-way digital afterlife platform	The use of AI algorithms to create a representation of the deceased	Users leave their data with a service such as Eternime so that their loved ones can talk to them after death	Bassett (2022) Savin-Baden, (2022).

To date there is little, if any, government intervention or legislation that deals with issues relating to the dead and the digital, leaving families faced with conundrums about what to do with digital remains. Furthermore, the growth of the dead on social media sites is having an increasing impact on climate, with their remains stored in large servers and data centres around the world. The development of digital afterlife creations, whether griefbots or digital immortals also will have an impact on the living. Thus, death and AI remain an increasing ethical concern, for which, as yet, there seems to be little action or planned resolution.

Conclusion

This chapter has examined a range of concerns about death, ethics and AI, but in many ways, this is only a brief overview of the many, varied and complex issues. It is clear from the ongoing research in this area that there are concerns about left-behind identities on social media and the general lack of discussion about ethics in relation to death and AI.

At the end of this text, it is clear that AI remains a deeply ethical concern, and perhaps what is more of a concern is that there appear to be few realistic solutions for tackling this. The final few pages of this book provide a suggested strategy for action which might begin some of the work for creating a realistic and ethical response to AI.

Reflection and strategy for realistic and responsible AI

The final few pages of this book provide a suggested strategy for action which might begin some of the work for creating a realistic and ethical response to AI.

The challenges

As we close this text, new and possibly worrying changes continue to develop. Reports of teachers using AI for marking in the UK are in the news, and the growth of the use of AI in drone warfare continues apace. It would be easy to finish on a negative note and suggest that AI poisoning and avoiding AI at all costs is the only responsible way ahead. The reality is that AI, of whatever kind, is here to stay. In practice what is needed is a roadmap for managing it now and in the future. To date, roadmaps or blueprints are not something that businesses, corporations or governments have provided. As mentioned in Chapter 7, the recent European Artificial Intelligence Act in 2024 (European Parliament & Council of the European Union, 2024) aimed to regulate AI systems, eliminate obstacles to trade and protect society against the adverse effects of AI – this may be the beginning of a process, but it will not be enacted for a few years to come.

However, recent statements by politicians (Bannerman, 2025) show little intention of regulating the theft of art. Therefore, the development of AI models free of stolen art seems unlikely unless it is explicitly made by the art community. In this regard, it seems like AI poisoning is the only option. However, one path forward seems to be to commission dedicated artists and writers to create the art needed to go into models. Providing the development of AI models to the artists will also likely ensure better models and more pro-art-focussed models such as automating sections of art that are tedious or difficult.

The future

We believe that what is likely to occur is an extensive proliferation of AI and, following a few years behind it, the legislation to regulate it as the negative consequences of the technology become realised. We have no idea what the AI landscape will look like in a few years, given that Large Language Models (LLMs) have completely changed the landscape on major issues such as the way education has been completely upended with students utilising ChatGPT for assignments. However, it is likely that, given AI integration is likely to persist over time, we are likely to see a significant advancement in healthcare, engineering and other industries, provided that the technology is integrated successfully and with a significant focus on accuracy. If the technology is faulty, then we are likely to see issues like the Theranos scandal, where a seemingly revolutionary blood testing technology turned out to be fraudulent. Therefore, these advancements must come with significant oversight and legislative restrictions to ensure that the technology is accurate and actively beneficial to humanity.

Suggestions for ethical and responsible AI from 2026 to 2030

These advancements will come at a significant cost, as discussed in Chapters 5, 6, and 9, and therefore, we are likely to see a massive energy rush, possibly culminating in a global energy crisis. Therefore, certain preventative measures should be implemented to prevent this as well as ensure that AI companies are regulating their energy consumption as well as committing to climate targets. We therefore suggest ideas for the creation of the realistic and ethical use of AI.

Effective global laws

What is needed are laws that sanction countries of unethical use of AI, which include:

- Standardise procedures for measuring the environmental impact of AI
- Regulate AI companies so that they must disclose the impact of their AI
- Increase AI efficiency so that less energy is used
- Use renewable energy to power data centres
- Implement policies on AI to meet environmental regulations
- Ensure that AI is created ethically and morally

Climate change

Realistic and effective targets are required to ensure green businesses, universities and colleges, such as the use of solar and wind power and the ethical sourcing of goods and hardware, as well as:

- The reduction of greenhouse gases through innovations such as smart cities and smart buildings
- Better weather forecasting may improve the ability to gain sustainable energy
- The creation of new energy sources such as nuclear fusion
- Scheduling of energy use so that is more effective and sustainable
- Changing transport to meet demand and changing freight routing

To date, the government response is highly mixed, with Germany and EU making good progress towards regulating AI to reduce climate change, while China has made some steps, and the United States is lagging badly behind.

Realistic stance towards carbon reduction

It is notable that carbon reduction effects of AI only exist in high-income and high-carbon-emission countries – which would seem to introduce questions about whether the use of AI for reducing carbon emission in such countries can really be effective in mitigating the high AI use in the first place.

Algorithmic ethics

In order that analysts can determine if an algorithm is illegal or unethical. For example, there needs to be improvement in the process of data collection such as contacting artists for using their work and better notification to those under CCTV that the system collects data and/or uses facial recognition technology. Overall, there needs to be a clear improvement to data collection, tools and resources addressing algorithmic bias.

Effective training in schools, colleges and universities about the use of Large Language Models

The use of ChatGPT and Llama (another opensource AI model), particularly throughout the education sector, means a significant rise in cheating and academic misconduct due to the usage of LLMs in essays and homework. Strict policies need to be implemented to mitigate misconduct.

High penalties for use of illegal images

While it is illegal for AI image generators to circulate explicit photos of minors or celebrities, there are little to no restrictions on this in the generators, and the generators often carry only a slight warning concerning their problematic content. It is also vital that the use of AI-generated images and graphics is reduced, if not banned.

Ethical use of drones and robots

The main concern here is that of accountability, more specifically who has it? The military that deploys it? The contractor that deploys it? The general who signs off on the drone deployment? The politicians that supported the bill through Parliament? Legislation needs to be in place world-wide to deal with this.

Personal responsibility

Few people realise the impact of their use of AI on the environment. The negligence and ignorance of the users is problematic as it will allow Big Tech corporations to have almost unchecked powers and influence regarding AI. Furthermore, as the users are unaware of the damage that they do, they will be less mindful of the way that AI is used. Therefore, there needs to be significant education around the dangers of AI technology.

Digital death trends

There needs to be recognition of and education about the impact of digital afterlife creation on the bereaved as well as understanding about the environmental impact of the cost of those left behind on Facebook, X and other social media sites. Furthermore, there needs to be an increase in education for the end of digital life and increased ease for closing accounts.

Conclusion: the hope

The major hope with AI is the utilisation of it within the medical sector as well as the automation of the tasks we find boring or monotonous, such as checking references. Other options include automatic vacuum cleaners which use pathfinding to clean the floor and the idea of Alexa to replace the classic encyclopaedia, as well as various other household features, such as smart fridges and heating systems, yet these are far from widespread or fully implemented within households. In terms of the medical sector, AI has made major strides with research as well as the detection of various health conditions and diseases, and the future seems very hopeful. Such strides include cancer detection, reading of radiology images, predicting patient outcomes and robotic surgery.

Glossary

AIML: Artificial intelligence markup language, a common language used to create chatbots and/or conversational interfaces.

Algorithms: a series of instructions or procedures that collect and encode data and based on specific calculations to transform data into output about a desired outcome.

Analytics: the computational analysis of statistics or data which locates meaningful patterns that may be sued in decision making

Artificial intelligence: broadly three main types.

1. Deep AI which involves creating virtual humans and digital immortals with sentience: the ability to feel, perceive or experience.
2. Pragmatic AI where machines are used for tasks such as mining large data sets or automating complex or risky tasks.
3. Marketing AI where companies use algorithms to *anticipate the customer's next move and improve the customer journey.*

Augmented reality: the live view of a physical world environment whose elements are merged with computer imagery; thus, it places emphasis on the physical, so that information from virtual space is accessed from within physical space. An example of this would be the projection of digital 3D objects into a physical space.

Augmented virtuality: this is where the virtual space is augmented with aspects from physical space, so there is a sense of overlay between the two spaces.

Autonomous agent: a software program which operates autonomously according to a particular set of rules. Sometimes, it is used synonymously with chatbots but also often used to describe the non-human agents within a simulation or decision-making system.

Avatar: the bodily manifestation of one's self or a virtual human in the context of a 3D virtual world, or even as a 2D image within a text-chat system.

Bluesky: a social media platform that was developed as an alternative to X (Twitter)

Catfishing: setting up a fake online identity for the purpose of tracking or manipulating someone

Chatbots: software programs which attempt to mimic human conversation when communicating with another (usually human) user. The Turing Test is a standard test of the maturity of chatbot technology. Chatbots may also be used to control 3D avatars within a virtual world, 2D avatars on a web site or exist as participants within text-only environments such as chat rooms. Chatbots are conversational agents which support a wide range of natural language and extended conversations, rather than just question and answer or command and response.

ChatGPT: a Generative AI model that can generate high-quality text, images and other content based on the data they were trained on; it uses algorithms to create new content, including audio, code, images, text, simulations and videos.

Clickbait: the creation of links designed to attract attention and encourage the user to click to view the content, which is often deceptive or sensational. Teasers are texts or links that provide sufficient information to make readers curious, so they click to get to the content

Computer Vision: the technical approaches which enable computers to see and understand imagery and video.

Conversational Agents: Computer programs which use a natural language rather than a command line (or other) interface to a system, database or program.

Co-Pilot: this is a Generative AI chatbot, based on Large Language Models and developed by Microsoft

Data Scraping: the act of extracting/collecting data from websites, documents, spreadsheets, databases, etc. Sometimes, carried out with automated tools.

Death Tech: the use of technology connected with death and bereavement and is a means of creating or retaining some form of digital afterlife.

DeepFakes: these are images or videos that have been edited to resemble real people using AI

DeepSeek: a free Generative AI model developed by Liang Wenfeng, a hedge fund entrepreneur in China

Digital afterlife: the continuation of an active or passive digital presence after death (Savin-Baden et al., 2017).

Digital fluency: the ability to shift easily between and across digital media, often unconsciously, with a sense of understanding of the value and possibilities of their use and function.

Digital human: a software version of a human which is typically focused on the size and shape of the human for ergonomic research purposes.

Digital inequalities: those inequalities such as skills, use and access that affect people's ability to exploit technology to support their life chances, as well as to grow and learn

Digital legacy: Digital assets left behind after death.

Digital literacy: is seen as the ability to assemble knowledge, evaluate information, search and navigate as well as locating, organising, understanding and evaluating information using digital technology.

Digital traces - the traces, or digital footprints, left behind by interaction with digital media. These tend to be of two types: intentional digital traces – emails, texts and blog posts; and unintentional digital traces – records of website searches and logs of movements

GitHub: a developer platform that allows developers to create, store, manage and share their code. A Git is a distributed version control system that tracks versions of files; it is often used to control source code by programmers who are developing software collaboratively.

Hallucination: a response generated by an AI which is either false or misleading information.

Lateral surveillance: the use of surveillance tools by individuals, rather than by agents of institutions public or private, to keep track of one another.

Learning context: the interplay of all the values, beliefs, relationships, frameworks and external structures that operate within a given learning environment.

Liquid surveillance: the idea that through data flows and regimes of in/visibility everyone is being targeted and sorted.

Llama (Large Language Model Meta AI): is a family of Large Language Models released by Meta AI and trained on only publicly available information

Machine Learning: a computer system that can learn to make decisions based on the examination of past inputs and results, so that its future decisions optimise some parameter such as facial recognition

Mixed reality: is seen as a method for integrating virtual and physical spaces much more closely so that physical and digital objects co-exist and interact in real time.

(The) Postdigital: stance which merges the old and the new; it is not seen as an event or temporal position, rather it is a critical perspective, a philosophy that can be summarised as a collection of stances.

Price-Gouging: the act of increasing the price of a good, service or commodity to a higher price than reasonable, often due to increased demand.

Scalping: a short-term trading strategy in cryptocurrency where traders aim to make numerous small profits by capitalising on minor price movements within a short timeframe.

Second Life: a persistent, shared, multi-user 3D virtual world launched in 2003 by Linden Lab. Residents (in the form of self-designed avatars) interact with each other and can learn, socialise, participate in activities and buy and sell items with one another, without any constraints of game play.

Speech Recognition: the techniques and technology which enable a computer to recognise and understand human speech.

Text chat: the means of communicating by text message and specifically in immersive virtual worlds by typing a response to another avatar in-world rather than using voice. Text chat may be private, public or in a closed group.

Text-to-Speech: the techniques and technology which can convert text (typically generated by a computer, but possibly also scanned from a page of text) to audible speech.

Transformative technologies: technologies that transform society and indeed the world in some way, such as the wheel and solar energy.

Virtual humanoids: Simple virtual humans which present, to a limited degree, as human and which may reflect some of the behaviour, emotion, thinking, autonomy and interaction of a physical human.

Virtual humans: Software programs which present as human and which have behaviour, emotion, thinking, autonomy and interaction modelled on physical humans.

Virtual reality: a simulated computer environment in either realistic or imaginary world. Most virtual reality emphasises immersion so that the user suspends belief and accepts it as a real environment and uses a head-mounted display to enhance this.

Voice recognition: the techniques and technology which enable a computer to recognise speech as belonging to a particular individual.

References

Ade-Ibijola, A., & Okonkwo, C. (2023). Artificial intelligence in Africa: Emerging challenges. In D. O. Eke., K. Wakunuma, & S. Akintoye, S. (Eds.), *Responsible AI in Africa: Challenges and opportunities* (pp. 101–117). Springer International Publishing.

Agarwal, N., Moehring, A., Rajpurkar, P., & Salz, T. (2023). *Combining human expertise with artificial intelligence: Experimental evidence from radiology.* SSRN Electronic Journal. https://doi.org/10.2139/ssrn.4505053

Ahn, M., Brohan, A., Brown, N., Chebotar, Y., Cortes, O., David, B., ... & Zeng, A. (2022). Do as I can, not as I say: Grounding language in robotic affordances. *arXiv preprint arXiv:2204.01691.*

Aibin, M. (2024, July 8). *Baeldung on computer science: ChatGPT, large language models, and power consumption.* Baeldung. https://www.baeldung.com/cs/chatgpt-large-language-models-power-consumption

Alkhalil, A., Abdallah, M. A. E., Alogali, A., & Aljaloud, A. (2021). Applying big data analytics in higher education: A systematic mapping study. *International Journal of Information and Communication Technology Education (IJICTE), 17*(3), 29–51.

Al-Sibai, N. (2024, July 4). *Researcher studying married men with AI girlfriends.* Futurism. https://futurism.com/researcher-married-men-ai-girlfriends

AMD. (2024, August 20). *Environmental sustainability.* AMD. https://www.amd.com/en/corporate/corporate-responsibility/environmental-sustainability.html

Anarumo, A. (Director). (2023). *Prom Pact* [TV film]. Disney+.

Anderson, M. L. (2003). Embodied cognition: A field guide. *Artificial Intelligence, 149*(1), 91–130.

Anguiano, D. (2023, December 6). Sag-Aftra union ratifies strike-ending contract with Hollywood studios. *The Guardian.* https://www.theguardian.com/culture/2023/dec/05/sag-aftra-union-ratifies-contract-hollywood-studios

Angwin, J., Larson, J., Mattu, S., & Kirchner, L. (2016). Machine Bias: there's software used across the country to predict future criminals. And it's biased against blacks. ProPublica. https://www.propublica.org/article/machine-bias-risk-assessments-in-criminal-sentencing

Apate.AI. (2024). *Apate.AI.* https://www.apate.ai/

AU10TIX. (2024, November 27). *The Q3 2024 Global Identity Fraud Report.* https://www.au10tix.com/landing/the-q3-2024-global-identity-fraud-report/

Australian Associated Press. (2023, September 27). Tesla lithium battery fire hits 'landmark' Queensland energy project, sparking political debate. *The Guardian.* https://www.theguardian.com/australia-news/2023/sep/27/tesla-lithium-battery-fire-bouldercombe-energy-storage-site-project-rockhampton

Badham, J. (Director). (1983). *WarGames* [Film]. United International Pictures.

Baker, R. S., & Hawn, A. (2022). Algorithmic bias in education. *International Journal of Artificial Intelligence in Education, 32*, 1052–1092. https://doi.org/10.1007/s40593-021-00285-9

Bannerman, L. (2025, May 23). Nick Clegg: Artists' demands over copyright are unworkable. *The Times*. https://www.thetimes.com/uk/technology-uk/article/nick-clegg-work-train-ai-l6xl5djb7

Barnett, R. (1994). *The limits of competence: Knowledge, higher education and society*. Open University Press.

Bassett, D. J. (2015). Who wants to live forever? Living, dying and grieving in our digital society. *Social Sciences, 4*(4), 1127–1139.

Bassett, D. J. (2022). *The creation and inheritance of digital afterlives*. Palgrave.

Bauwens, M. (2005). The political economy of peer production. In A. Kroker & M. Kroker (Eds.), *1000 days of theory*. https://journals.uvic.ca/index.php/ctheory/article/view/14464

Bender, E. M., & Koller, A. (2020, July). Climbing towards NLU: On meaning, form, and understanding in the age of data. In *Proceedings of the 58th annual meeting of the association for computational linguistics* (pp. 5185–5198). Association for Computational Linguistics.

Berman, R. (1995). Star Trek: Voyager [TV series]. CBS Television.

Besser, A., Morse, T., & Zeigler-Hill, V. (2023). Who wants to (digitally) live forever? The connections that narcissism has with motives for digital immortality and the desire for digital avatars. *International Journal of Environmental Research and Public Health, 20*(17), 6632.

Big Brother Watch. (2019). *FaceOff*. May. https://bigbrotherwatch.org.uk/campaigns/stop-facial-recognition/

Billingham, P., & Parr, T. (2020). Enforcing social norms: The morality of public shaming. *European Journal of Philosophy, 28*(4), 997–1016.

Block, N. (1981). Psychologism and behaviorism. *The Philosophical Review, 90*(1), 5–43.

Bloomberg News. (2024, October). China clean energy sector eyes $100 billion in overseas spending. *Bloomberg*. https://www.bloomberg.com/news/articles/2024-10-01/china-clean-energy-sector-eyes-100-billion-in-overseas-spending

Brackley, J., & Vincent, S. (Creators). (2015–2018). *Humans* [TV series]. Kudos Film and Television; AMC Studios.

Bradeško, L., & Mladenić, D. (2012). A survey of chatbot systems through a Loebner Prize competition. In *Proceedings of the Slovenian Language Technologies Society Eighth Conference of Language Technologies* (Vol. 2, pp. 34–37). Jožef Stefan Institute. https://nl.ijs.si/isjt12/proceedings/isjt2012_06.pdf

Bronte, C. (1847) *Jane Eyre*. Penguin.

Brown, M. (2021). Ethics, character, and community: Moral formation and modelling the human. In M. Savin-Baden (Ed.), *Postdigital humans: Transitions, transformations and transcendence* (pp. 125–140). Springer International Publishing. https://doi.org/10.1007/978-3-030-65592-1_8

Brown, M. (2020, November 3). SpaceX Mars city: Legal experts respond to "gibberish" free planet claim. *Inverse*. https://www.inverse.com/innovation/spacex-mars-city-legal

Brubaker, J. R., & Callison-Burch, V. (2016, May). Legacy contact: Designing and implementing post-mortem stewardship at Facebook. In *Proceedings of the 2016 CHI conference on human factors in computing systems* (pp. 2908–2919). San Jose, CA. https://doi.org/10.1145/2858036.2858254

Buben, A. (2015). Technology of the dead: Objects of loving remembrance or replaceable resources? *Philosophical Papers, 44* (1), 15–37. https://doi.org/10.1080/05568641.2015.1014538

Bundesumweltministeriums. (2024, April 26). *KI-Leuchttürme für Umwelt, Klima, Natur und Ressourcen – BMUV – Förderprogramm.* BMUKN. https://www.bmuv.de/programm/ki-leuchttuerme-fuer-umwelt-klima-natur-und-ressourcen

Buolamwini, J., & Gebru, T. (2018, January). Gender shades: Intersectional accuracy disparities in commercial gender classification. In *Conference on Fairness, Accountability and Transparency* (pp. 77–91). PMLR.

Burden, D. J. (2009). Deploying embodied AI into virtual worlds. *Knowledge-Based Systems, 22*(7), 540–544.

Burden, D. J., Savin-Baden, M., & Bhakta, R. (2016). Covert implementations of the Turing test: A more level playing field? In M. Bramer & M. Petridis (Eds.), *Research and development in intelligent systems XXXIII: Incorporating applications and innovations in intelligent systems XXIV 33* (pp. 195–207). Springer.

Burden, D., & Savin-Baden, M (2019). *Virtual humans: Today and tomorrow.* CRC Press.

Burden, D. (2020). Building a digital immortal. In M. Savin-Baden & V. Mason-Robbie (Eds.), *Digital afterlife* (pp. 143–160). Chapman and Hall/CRC.

Calhoun, A. W., Pian-Smith, M. C. M., Truog, R. D., Gaba, D. M., & Meyer, E. C. (2015). Deception and simulation education: Issues, concepts, and commentary. *Simulation in Healthcare, 10*(3), 163–169.

Calma, J. (2024, May 20). Google plans to reuse heat after expanding a data center for AI. *The Verge.* https://www.theverge.com/2024/5/20/24160788/google-ai-data-center-expansion-heat-recovery

Cameron, J. (Director). (1984). *The Terminator* [Film]. Orion Pictures.

capable.design. (2025). *Design for awareness | Cap_able design.* Cap_able. https://www.capable.design/

Carroll, M. (2024, October 24). Mother says son killed himself because of "hypersexualised" and "frighteningly realistic" AI chatbot in new lawsuit. *Sky News.* https://news.sky.com/story/mother-says-son-killed-himself-because-of-hypersexualised-and-frighteningly-realistic-ai-chatbot-in-new-lawsuit-13240210

Cassidy, C. (2023, January 17). Lecturer detects bot-use in one fifth of assessments as concerns mount over AI in exams. *The Guardian.* https://www.theguardian.com/australianews/2023/jan/17/lecturer-detects-bot-use-in-one-fifthof-assessments-as-concerns-mount-over-ai-in-exams

Chan, R. (2019, December). *Activism spreads in open source community, as tech workers protest ICE.* Business Insider. https://www.businessinsider.com/open-source-activism-ice-protests-open-source-initiative-2019-11

Chapman, B. (2018, January 9). Nearly two-thirds of adult don't have will, research finds. *Independent.* January 9. Available https://www.independent.co.uk/news/business/news/nearly-two-thirds-uk-adults-don-t-have-will-research-finds-a8148316.html

Chaman, B. (Writer & Director). (2019). *The Capture* [TV series]. BBC.

Chen, P., Gao, J., Ji, Z., Liang, H., & Peng, Y. (2022). Do artificial intelligence applications affect carbon emission performance? Evidence from panel data analysis of Chinese cities. *Energies, 15*(15), 5730. https://doi.org/10.3390/en15155730

Chikwem, H. (2024, June 20). The rise of deceased social media accounts. *Techloy.* https://www.techloy.com/the-rise-of-deceased-social-media-accounts/amp/

Chithaluru, P., Al-Turjman, F., Kumar, M., & Stephan, T. (2023). Computational-intelligence-inspired adaptive opportunistic clustering approach for industrial IoT networks. *IEEE Internet of Things Journal, 10*(9), 7884–7892.

Chow, K.-H., Hu, S., Huang, T., & Liu, L. (2024). Personalized privacy protection mask against unauthorized facial recognition. *Lecture Notes in Computer Science*, 434–450. https://doi.org/10.1007/978-3-031-73007-8_25

Cihon, P., Maas, M. M., & Kemp, L. (2021). Corporate governance of AI in the public interest. *Information*, 12(7), 275. https://www.mdpi.com/2078-2489/12/7/275

Cohen, E., & Spector, S. (2020). Transhumanism and cosmic travel. *Tourism Recreation Research*, 45(2), 176–184.

Collins, E. (2018). Punishing risk. *Georgetown. Law Journal*, 107(1) 57–108.

Common Crawl. (2024). *Common Crawl*. https://commoncrawl.org/

Corey, J., Ostby, H., & Fergus, M. (2015, December 14). *The Expanse* [TV series]. Amazon Prime; Syfy.

Couldry, N., & Mejias, U. A. (2019). Data colonialism: Rethinking big data's relation to the contemporary subject. *Television & New Media*, 20(4), 336–349.

Crawford, K. (2021). *The atlas of AI: Power, politics, and the planetary costs of artificial intelligence*. Yale University Press.

Creamer, E. (2025, April 4). US authors' copyright lawsuits against OpenAI and Microsoft combined in New York with newspaper actions. *The Guardian*. https://www.theguardian.com/books/2025/apr/04/us-authors-copyright-lawsuits-against-openai-and-microsoft-combined-in-new-york-with-newspaper-actions

da Silva, J. (2024, October 15). Google turns to nuclear to power AI data centres. *BBC News*. https://www.bbc.co.uk/news/articles/c748gn94k95o

Dafermos, G., & Söderberg, J. (2009). The hacker movement as a continuation of labour struggle. *Capital & Class*, 33(1), 53–73.

Daly, M., & Borenstein, S. (2025, January 20). Trump says he's withdrawing the US from the Paris climate agreement again. *AP News*. https://apnews.com/article/trump-paris-agreement-climate-change-788907bb89fe307a964be757313cdfb0

Dastin, J., & Nellis, S. (2023). Focus: For tech giants, AI like Bing and Bard poses billion-dollar search problem. *Reuters*. https://www.reuters.com

Davids, N. (2021). Academic freedom and the fallacy of a post-truth era. *Educational Philosophy and Theory*, 53(11), 1183–1193.

Daws, R. (2019, December 12). *Palantir took over Project Maven defense contract after Google backed out*. AI News. https://www.artificialintelligence-news.com/news/palantir-project-maven-defense-contract-google-out/

Deeley, M. (Producer) & Scott, R. (Director), (1982) *Blade Runner* [Motion Picture]. Warner Bros.

de Godoy, J., Otrel-Cass, K., & Toft, K. H. (2021). Transformations of trust in society: A systematic review of how access to big data in energy systems challenges Scandinavian culture. *Energy and AI*, 5, 100079. https://doi.org/10.1016/j.egyai.2021.100079

De Vries, A. (2023). The growing energy footprint of artificial intelligence. *Joule*, 7(10), 2191–2194. https://doi.org/10.1016/j.joule.2023.09.009

Dejevsky, M. (2023, September 11). The trouble with supermarket checkouts. *The Spectator*. https://www.spectator.co.uk/article/the-trouble-with-supermarket-self-checkouts/

Delcid, N. (2022). Is Google's AI sentient? Stanford AI experts say that's 'pure clickbait'. *The Stanford Daily*.

Derrida, J. (1997). *Of grammatology. (G. Spivak trans.)*. Johns Hopkins University Press.

Dickens, C. (1843) *A Christmas carol*. Chapman and Hall.

Dimensions in Testimony: The Interviewees. *USC Shoah Foundation*. Available https://sfi.usc.edu/dit/interviewees

Dincer, A., Yeşilyurt, S., Noels, K. A., & Vargas Lascano, D. I. (2019). Self-determination and classroom engagement of EFL learners: A mixed-methods study of the self-system model of motivational development. *Sage Open, 9*(2), 2158244019853913

Donhauser, J. (2019). Environmental robot virtues and ecological justice. *Journal of Human Rights and the Environment, 10*(2), 176–192. https://doi.org/10.4337/jhre.2019.02.04

Donhauser, J. (2024). AI and the environment. In D. Gunkel (Ed.), *Handbook on the ethics of artificial intelligence* (pp. 187–204). Edward Elgar Publishing.

Driess, D., Xia, F., Sajjadi, M. S., Lynch, C., Chowdhery, A., Wahid, A., Tompson, J., Vuong, Q., Yu, T., Huang, W., & Chebotar, Y. (2023). Palm-e: An embodied multimodal language model. Available https://openreview.net/pdf?id=VTpHpqM3Cf

D'Silva, J., & Turner, J. (Eds.). (2012). *Animals, ethics and trade: The challenge of animal sentience*. Routledge.

D'Souza, J. (2025, May 12). *Duolingo statistics by revenue, users and facts (2025)*. Electro IQ. https://electroiq.com/stats/duolingo-statistics/

Ehrlich, M. (2017). *Navy turns to NPS to develop Sea Hunter's potential future missions*. Naval Postgraduate School. https://nps.edu/-/navy-turns-to-nps-to-develop-sea-hunter-s-potential-future-missio-1

Eliot, T. S. (1910/1996). The Triumph of Bullshit. *Inventions of the March Hare: Poems 1909–1917*. Harcourt.

Encyclopedia Britannica. (2024). *How Henry Ford's assembly line revolutionized factory production*. https://www.britannica.com/video/who-was-Henry-Ford/-259364

Ensign, D., Friedler, S. A., Neville, S., Scheidegger, C., & Venkatasubramanian, S. (2018, January). Runaway feedback loops in predictive policing. In *Conference on Fairness, Accountability and Transparency* (pp. 160–171). PMLR.

Epstein, R., Roberts, G., & Beber, G. (Eds.). (2009). *Parsing the Turing test*. Springer.

Euronews. (2024, October 17). Amazon joins Google in taking the nuclear option to power data centres. *Euronews*. https://www.euronews.com/business/2024/10/17/amazon-follows-google-in-taking-the-nuclear-option-to-power-data-centres

European Parliament & Council of the European Union. (2024). *Regulation (EU) 2024/1689 of the European Parliament and of the Council of 13 June 2024 laying down harmonised rules on artificial intelligence and amending Regulations (EC)*. Official Journal of the European Union, L 2024/1689.

European Union. (2016). *Regulation (EU) 2016/679 of the European Parliament and of the Council of 27 April 2016 on the protection of natural persons with regard to the processing of personal data and on the free movement of such data (General Data Protection Regulation)*. Official Journal of the European Union. https://eur-lex.europa.eu/eli/reg/2016/679/oj

f4mi. (2025, January 22). *Poisoning AI with ".ass" subtitles* [Video]. YouTube. https://www.youtube.com/watch?v=NEDFUjqA1s8

Fan, F., Xiong, J., & Wang, G. (2020). On interpretability of artificial neural networks. *arXiv*. https://arxiv.org/abs/2001.02522

Fang, R. (2022, March 29). *Google is using us to train their self driving cars*. ILLUMINATION. https://medium.com/illumination/google-is-using-us-to-train-their-self-driving-cars-88d024a4fa27

Feige, K. (Producer) & Whedon, J. (Directors (2015) *Avengers: Age of Ultron* [Motion Picture]. Marvel Studios.

Fitzpatrick, K. (2021). *Generous thinking: A radical approach to saving the university*. Johns Hopkins University Press.

Friedman, B., & Nissenbaum, H. (1996). Bias in computer systems. *ACM Transactions on information systems (TOIS)*, *14*(3), 330–347.

Foucault, M. (2008). Of other spaces*(1967). In M. Dehaene & L. De Cauter (Eds.), *Heterotopia and the city* (pp. 25–42). Routledge.

Foucault, M., & Miskowiec, J. (1986). Of other spaces. *Diacritics*, *16*(1), 22–27.

Frude, N., & Jandric, P. (2015). The intimate machine – 30 years on. *E-Learning and Digital Media*, *12*(3–4), 410–424.

Gagne, Y. (2023, October 8). Why Gen Z is obsessed with the Duolingo Owl. *Fast Company*. https://www.fastcompany.com/90963949/duolingo-duo-owl-gen-z-obsessed

Gailhofer, P., Herold, A., Schemmel, J., Scherf, C.-S., Urrutia, C., Köhler, A., & Braungardt, S. (2021). *The role of artificial intelligence in the European Green Deal*. https://www.europarl.europa.eu/RegData/etudes/STUD/2021/662906/IPOL_STU(2021)662906_EN.pdf

Garcia, M. (2016). Racist in the machine. *World Policy Journal*, *33*(4), 111–117.

Germain, M.-L., & Grenier, R. S. (2022). *Expertise at work*. Palgrave Macmillan.

German Federal Government. (2020, December). *Artificial intelligence strategy of the German Federal Government: 2020 update*. https://www.ki-strategie-deutschland.de/files/downloads/Fortschreibung_KI-Strategie_engl.pdf

Gergen, K. J. (2009). *Relational being: Beyond self and community*. Oxford University Press.

Ghosh, P. (2018, October 31). Paralysed men walk again with spinal implant. *BBC News*. https://www.bbc.co.uk/news/health-46043924

Gilroy, T. (Writer), & Edwards, G. (Director). (2016). *Rogue one: A Star Wars story* [Film]. Lucasfilm; Walt Disney Studios Motion Pictures.

Global Institute for National Capacity. (2024, November 17). *China's National AI Strategy*. https://www.ginc.org/chinas-national-ai-strategy/

Goldman, L. M. (2015). Trending now: the use of social media websites in public shaming punishments. *American Criminal Law Review*, *52*, 415.

Gooding, M. (2025, January 20). Legislation proposed in Virginia to regulate data centers, ease grid concerns. *Datacenter Dynamics*. https://www.datacenterdynamics.com/en/news/virginia-data-center-laws/

Grace, K., Salvatier, J., Dafoe, A., Zhang, B., & Evans, O. (2018). When will AI exceed human performance? Evidence from AI experts. *Journal of Artificial Intelligence Research*, *62*, 729–754. https://doi.org/10.1613/jair.1.11222

Grandinetti, J. (2020). Reaminator: Haunted data, streaming media, and subjectivity in J. Grandinetti., T. DeAtley, & J. Bruinsma. The dead speak: Big data and digitally mediated death. Panel Presented at *Aoir 2020: The 21th Annual Conference of the Association of Internet Researchers*. Virtual Event: AoIR. Available online https://spir.aoir.org.

Greenpeace. (2020). *Oil in the cloud: How tech companies are helping big oil profit from climate destruction*. Greenpeace. https://www.greenpeace.org/usa/reports/oil-in-the-cloud/

Grzankowski, A. (2025, February 5). *Benchmarking intelligence better - Ethics in AI Lunchtime Research Seminars*. University of Oxford.

Guanio-Uluru, L. (2016). War, games, and the ethics of fiction. *Game Studies*, *16*(2). https://hdl.handle.net/10852/58768

Harari, Y. N. (2015). *Homo Deus*. Harvill Secker.

Harari, Y. N. (2017). Life 3.0 by Max Tegmark review – we are ignoring the AI apocalypse. *The Guardian*. Available online https://www.theguardian.com/books/2017/sep/22/life-30-max-tegmark-review

Hayes, S. (2021). *Postdigital positionality: Developing powerful inclusive narratives for learning, teaching, research and policy in Higher Education*. Brill.

Hayes, S., Connor, S., Johnson, M., & Jopling, M. (Eds.). (2023). *Human data interaction, disadvantage and skills in the community: Enabling cross-sector environments for post-digital inclusion*. Springer. https://doi.org/10.1007/978-3-031-3187

Heikkilä, M. (2024, June 14). How to opt out of Meta's AI training. *MIT Technology Review*. https://www.technologyreview.com/2024/06/14/1093789/how-to-opt-out-of-meta-ai-training/

Hern, A. (2023). Amazon's Alexa could turn dead loved ones' voices into digital assistant. Available online: https://www.theguardian.com/technology/2022/jun/23/amazon-alexa-could-turn-dead-loved-ones-digital-assistant

Hess, K., & Waller, L. (2014). The digital pillory: Media shaming of 'ordinary' people for minor crimes. *Continuum, 28*(1), 101–111.

Hill, K. (2022). Wrongfully accused by an algorithm. In K. Martin (Ed.), *Ethics of data and analytics: Concepts and cases* (pp. 138–142). Auerbach Publications.

Hu, M. (2020). Cambridge Analytica's black box. *Big Data & Society, 7*(2), 2053951720938091.

Ibbetson, C. (2019). *What should happen to data and social media accounts when people die?* YouGov. https://yougov.co.uk/topics/society/articles-reports/2019/11/01/what-do-brits-want-happen-their-data-and-social-me

Jain, R., Triandis, H. C., & Weick, C. W. (2010). *Managing research, development and innovation: Managing the unmanageable* (Vol. 34). John Wiley & Sons.

Jain, P., Agarwal, P., Mathur, D., Singh, P., & Sharma, A. (2021). Evolution of cryogenics–A review on applications of cryogenics in medicine. *Materials Today: Proceedings, 47*, 3059–3063.

Jandrić, P., Knox, J., Besley, T., Ryberg, T., Suoranta, J., & Hayes, S. (2018). Postdigital science and education. In *Educational philosophy and theory* (Vol. 50, Issue 10, pp. 893–899). Taylor & Francis.

Jandrić, P. (2019). The postdigital challenge of critical media literacy. *The International Journal of Critical Media Literacy, 1*(1), 26–37.

Jecker, N. S., Sparrow, R., Lederman, Z., & Ho, A. (2024). Digital humans to combat loneliness and social isolation: ethics concerns and policy recommendations. *Hastings Center Report, 54*(1), 7–12.

Johnson, J. T. (2018). Religion and the human rights idea. *Journal of Religious Ethics, 46*(2), 379–398.

Johnson, J. (2024, March 18). *Jus ad bellum | Law | Britannica*. Britannica. https://www.britannica.com/topic/jus-ad-bellum

Kaack, L. H., Donti, P. L., Strubell, E., Kamiya, G., Creutzig, F., & Rolnick, D. (2022). Aligning artificial intelligence with climate change mitigation. *Nature Climate Change, 12*(6), 518–527. https://doi.org/10.1038/s41558-022-01377-7

Kan, M. (2021, November 19). Read it and weep: Here's how bad Nvidia GPU prices got in a single year. *PCMag*. https://www.pcmag.com/news/read-it-and-weep-heres-how-bad-nvidia-gpu-prices-got-in-a-single-year

Kant, I. (1870/1998). *Grundlegung zur Metaphysik der Sitten*, translated and edited by M. Gregor, *Groundwork of the Metaphysics of Morals*. Cambridge University Press.

Karas, R. S. (2017). Lockheed's latest Loyal Wingman test let unmanned F-16 chart its own path. *Inside the Air Force, 28*(15), 1–5. https://www.jstor.org/stable/26411586

Kasket, E. (2019). *All the ghosts in the machine: Illusions of immortality in the digital age.* Robinson.

Kasket, E. (2021). If death is the spectacle, big tech is the lens: How social media frame an age of 'spectacular death.' In M. H. Jacobsen (Ed.), *The age of spectacular death* (pp. 20–35). Routledge.

Khomami, N. AI used to create new and final Beatles song, says Paul McCartney. Available online: https://www.theguardian.com/music/2023/jun/13/ai-used-to-create-new-and-final-beatles-song-says-paul-mccartney

Kitchin, R. (2014). Big Data, new epistemologies and paradigm shifts. *Big Data & Society, 1*(1), 2053951714528481.

Kitchin, R., & McArdle, G. (2016). What makes big data, big data? Exploring the ontological characteristics of 26 datasets. *Big Data & Society, 3*(1), 2053951716631130.

Klare, B. F., Burge, M. J., Klontz, J. C., Bruegge, R. W. V., & Jain, A. K. (2012). Face recognition performance: Role of demographic information. *IEEE Transactions on Information Forensics and Security, 7*(6), 1789–1801.

Klass, D., Silverman, P. R., & Nickman, S. L. (Eds.). (1996). *Continuing bonds: New understandings of grief.* Taylor & Francis.

Klass, D. (2018). Prologue. In D. Klass & E. M. Steffen (Eds.), *Continuing bonds in bereavement: New directions for research and practice* (pp. xiii–xix). Routledge.

Kleeman, J. (2025, February 8). Elon Musk put a chip in this paralysed man's brain. Now he can move things with his mind. Should we be amazed - or terrified? *The Guardian.* https://www.theguardian.com/science/2025/feb/08/elon-musk-chip-paralysed-man-noland-arbaugh-chip-brain-neuralink

Koopman, S. (2025, April 14). EY reveals Brits embrace AI at home, not at work. *City A.M.* https://www.cityam.com/ey-report-shows-brits-embrace-ai-at-home-but-shun-it-at-work/

Kosove, A. A., Johnson, B., Cohen, K., Polvino, M. Marter, A., Valder, D., Ryder, A. (Producers) & Pfister, W. (Director), (2014) *Transcendence.* [Motion Picture]. Warner Bros.

Koutsouleris, N., Hauser, T. U., Skvortsova, V., & De Choudhury, M. (2022). From promise to practice: Towards the realisation of AI-informed mental health care. *The Lancet Digital Health, 4*(11), e829–e840 https://doi.org/10.1016/s2589-7500(22)00153-4

Kuo, T. C., Kuo, C. Y., & Chen, L. W. (2022). Assessing environmental impacts of nanoscale semiconductor manufacturing from the life cycle assessment perspective. *Resources, Conservation and Recycling, 182,* 106289. https://doi.org/10.1016/j.resconrec.2022.106289

Kwebbelkop. (2022, March 13). *Kwebbelkop AI 2.0* [Video]. YouTube. https://www.youtube.com/channel/UC3UanmVRKWJpHG8IVYkldVA

Lai, V. D., Ngo, N. T., Veyseh, A. P. B., Man, H., Dernoncourt, F., Bui, T., & Nguyen, T. H. (2023). Chatgpt beyond English: Towards a comprehensive evaluation of large language models in multilingual learning. *arXiv preprint arXiv:2304.05613.*

Lapuschkin, S., Wäldchen, S., Binder, A., Montavon, G., Samek, W., & Müller, K.-R. (2019). Unmasking Clever Hans predictors and assessing what machines really learn. *Nature Communications, 10*(1), 1–8. https://doi.org/10.1038/s41467-019-08987-4

Lawford, M. (2021, November 10). How much would you pay to come back from the dead? The starting price is £60k. *The Telegraph.* https://www.telegraph.co.uk/money/consumer-affairs/much-would-pay-come-back-dead-starting-price-60k/

Lawler, R. (2022, February 1). *Crisis text line stops sharing conversation data with AI company*. The Verge. https://www.theverge.com/2022/1/31/22906979/crisis-text-line-loris-ai-epic-privacy-mental-health

Lee, D. (2018). Why big tech pays poor Kenyans to teach self-driving cars. *BBC Technology*, 3 November. https://www.bbc.com/news/technology-46055595.

Legal500. (2023, June 23). *Open-source software – usage & its legal implications – Legal developments*. https://www.legal500.com/developments/thought-leadership/open-source-software-usage-its-legal-implications/

Le Ray, G., & Pinson, P. (2020). The ethical smart grid: Enabling a fruitful and long-lasting relationship between utilities and customers. *Energy Policy, 140,* 111258. https://doi.org/10.1016/j.enpol.2020.111258

Lenat, D., & Guha, R. V. (1989). *Building large knowledge-based systems: Representation and inference in the Cyc project*. Addison- Wesley Publishing Company

Li, P., Yang, J., Islam, M. A., & Ren, S. (2023). Making AI less "thirsty": Uncovering and addressing the secret water footprint of AI models. arXiv. https://arxiv.org/abs/2304.03271

Lian, Y., Tang, H., Xiang, M., & Dong, X. (2024). Public attitudes and sentiments toward ChatGPT in China: A text mining analysis based on social media. *Technology in Society, 76,* 102442.

Lippi, M., Pałka, P., Contissa, G., Lagioia, F., Micklitz, H.-W., Sartor, G., & Torroni, P. (2019). CLAUDETTE: An automated detector of potentially unfair clauses in online terms of service. *Artificial Intelligence and Law, 27*(2), 117–139. https://doi.org/10.1007/s10506-019-09243-2

Losh, E. (2009). Regulating violence in virtual worlds: Theorizing just war and defining war crimes in World of Warcraft. *Pacific Coast Philology, 44*(2), 159–172.

Loveys, K., Sagar, M., Pickering, I., & Broadbent, E. (2021). A digital human for delivering a remote loneliness and stress intervention to at-risk younger and older adults during the COVID-19 pandemic: randomized pilot trial. *JMIR mental health, 8*(11), e31586.

Lu, D. (2019) AI can predict if you'll die soon – but we've no idea how it works. *New Scientist*, November. Available https://www.newscientist.com/article/2222907-ai-can-predict-if-youll-die-soon-but-weve-no-idea-how-it-works/

Lu, M., & Qiu, J. L. (2022). Empowerment or warfare? Dark skin, AI camera, and Transsion's patent narratives. *Information, Communication & Society, 25*(6), 768–784.

Lucas, G. (Writer & Director). (1977). *This Episode IV – A New Hope* [Film]. Lucasfilm; 20th Century Fox.

Luccioni, S., Jernite, Y., & Strubell, E. (2024, June). Power hungry processing: Watts driving the cost of AI deployment? In *Proceedings of the 2024 ACM Conference on Fairness, Accountability, and Transparency* (pp. 85–99). ACM. https://doi.org/10.1145/3593013.3594033

Lund, A. (2024, November 22). *"Rogue One" technology used to bring Peter Cushing back to "Star Wars" is obsolete*. MovieWeb. https://movieweb.com/star-wars-rogue-one-cgi-peter-cushing-tarkin-technology-obsolete/

Lungariello, M. (2025, April 13). Ailing "Doctor Who" superfan spends fortune to recreate 97 lost episodes to see "complete" series before he dies. *New York Post*. https://nypost.com/2025/04/13/entertainment/ailing-doctor-who-superfan-spends-fortune-to-recreate-97-lost-episodes-to-see-complete-series-before-he-dies/

Maciel, C. (2011). Issues of the social web interaction project faced with afterlife digital legacy. In: *Proceedings of the 10th Brazilian Symposium on Human Factors in Computing Systems and the 5th Latin American Conference on Human-Computer Interaction* (pp. 3–12). ACM Press.

Maciel, C., & Pereira, V. (2013). *Digital legacy and interaction.* Springer.

MacFarlane, S. (2017, September 10). *The Orville* [TV series]. Fox; Hulu.

MacIntyre, A. (2016). *Ethics in the conflicts of modernity: An essay on desire, practical reasoning and narrative.* Cambridge University Press.

MacKenzie, A., & Bhatt, I. (2020). Lies, bullshit and fake news: Some epistemological concerns. In *Postdigital science and education* (Vol. 2, Issue 1, pp. 9–13). Springer International Publishing. https://doi.org/10.1007/s42438-018-0025-4

Mahajan, S. (2023). Artificial intelligence and its impacts on the society. *Contemporary Social Sciences, 32*(4), 135–151. https://doi.org/10.62047/css.2023.12.31.135

Mann, M. E. (2021). *The new climate war: The fight to take back our planet.* PublicAffairs.

Mateos-Garcia, J., Witherspoon, S., & Gabriel, J. (2024). Environmental impact. In I. Gabriel et al. (Eds.), *The ethics of advanced AI assistants.* arXiv. https://arxiv.org/abs/2404.16244

McCloskey, M., & Cohen, N. J. (1989). Catastrophic interference in connectionist networks: The sequential learning problem. In G. H. Bower (Ed.), *Psychology of learning and motivation* (Vol. 24, pp. 109–165). Academic Press. https://www.sciencedirect.com/science/article/abs/pii/S0079742108605368?via%3Dihub

Microsoft. (2024a). *Microsoft privacy statement – Microsoft privacy.* Microsoft.com. https://www.microsoft.com/en-us/privacy/privacystatement

Microsoft. (2024b, July 22). *Connected experiences and your content - Microsoft 365 Apps.* Microsoft.com. https://learn.microsoft.com/en-us/microsoft-365-apps/privacy/connected-experiences-content#connected-experiences-and-machine-learning

Milmo, D., & Courea, E. (2024, February 11). US and UK refuse to sign Paris summit declaration on 'inclusive' AI. *The Guardian.* https://www.theguardian.com/technology/2025/feb/11/us-uk-paris-ai-summit-artificial-intelligence-declaration

Minerva, F., & Giubilini, A. (2023). Is AI the future of mental healthcare? *Philosophia, 42*(3), 809–817. https://doi.org/10.1007/s11245-023-09932-3

Ministry of Justice. (2025, January 7). Government crackdown on explicit deepfakes. *GOV. UK.* https://www.gov.uk/government/news/government-crackdown-on-explicit-deepfakes

Moreno, A., & Redondo, T. (2016). Text analytics: The convergence of big data and artificial intelligence. *IJIMAI, 3*(6), 57–64.

Mori, M. (1970). Bukimi no tani [The uncanny valley]. *Energy, 7*(4), 33–35.

Morse, T. (2024). Digital necromancy: Users' perceptions of digital afterlife and posthumous communication technologies. *Information, Communication & Society, 27*(2), 240–256.

Mortier, R., Haddadi, H., Henderson, T., McAuley, D., & Crowcroft, J. (2014). Human data interaction: The human face of the data-driven society. *SSRN Electronic Journal.* https://doi.org/10.2139/ssrn.2508051

Mr.Rand0mStuf. (2024, January 9). *B1 Battle Droid sings "Wellerman"* | *AI cover* [Video]. YouTube. https://www.youtube.com/watch?v=mH_m6rbkM3E

Mullock, A., & Romanis, E. C. (2023). Cryopreservation and current legal problems: Seeking and selling immortality. *Journal of Law and the Biosciences, 10*(2), lsad028.

Munchetty, N. (2025, February 5). Scammers spread fake nude pictures of me on social media. *BBC News*. https://www.bbc.co.uk/news/articles/c1ezq8ll792o]

NaNoWriMo. (2025). *NaNoWriMo offers*. https://nanowrimo.org/offers

Nast, C. (2017). How a son made a chatbot of his dying dad. *WIRED*. Retrieved August 21, 2022, from https://www.wired.com/video/watch/how-a-son-made-a-chat-bot-of-his-dying-dad.

Nayar, J. (2021, August 12). Not so 'green' technology: The complicated legacy of rare earth mining. *Harvard International Review*. https://hir.harvard.edu/not-so-green-technology-the-complicated-legacy-of-rare-earth-mining/

Netsafe. (2017). *Netsafe's Re:Scam AI tool to waste scammers time*. https://netsafe.org.nz/rescam

Neumeister, L. (2025, April 4). An AI avatar tried to argue a case before a New York court. The judges weren't having it. *AP News*. https://apnews.com/article/artificial-intelligence-ai-courts-nyc-5c97cba3f3757d9ab3c2e5840127f765

Newton, C. (2016) Speak, Memory. Available https://www.theverge.com/a/luka-artificial-intelligence-memorial-roman-mazurenko-bot

Ng, E. (2020). No Grand pronouncements here...: Reflections on cancel culture and digital media participation. *Television & New Media*, 21(6), 621–627. https://doi.org/10.1177/1527476420918828

Noble, S. U. (2018). *Algorithms of oppression: How search engines reinforce racism*. New York University Press.

Note GPT. (2025). *AI podcast generator – Online free, no sign-up*. https://notegpt.io/ai-podcast-generator

NVIDIA. (2024). *Corporate sustainability*. NVIDIA. https://www.nvidia.com/en-us/sustainability/

nyuuzyou. (2025). *archiveofourown* [Dataset]. Hugging Face. https://huggingface.co/datasets/nyuuzyou/archiveofourown

O2. (2024). *O2 scam help | Meet dAIsy, the scam-fighting AI bot*. https://www.o2.co.uk/inspiration/the-drop/meet-daisy-the-scam-fighting-ai-bot

Obar, J. A., & Oeldorf-Hirsch, A. (2020). The biggest lie on the Internet: Ignoring the privacy policies and terms of service policies of social networking services. *Information, Communication & Society*, 23(1), 128–147. https://doi.org/10.1080/1369118x.2018.1486870

Olcott, E., & Criddle, C. (2025, January 29). OpenAI says it has evidence China's DeepSeek used its model to train competitor. *Financial Times*. https://www.ft.com/content/a0dfedd1-5255-4fa9-8ccc-1fe01de87ea6

Olesen, T. (2022). Whistleblowing in a time of digital (in) visibility: Towards a sociology of 'grey areas'. *Information, Communication & Society*, 25(2), 295–310.

Ollion, É., Shen, R., Macanovic, A., & Chatelain, A. (2024). The dangers of using proprietary LLMs for research. *Nature Machine Intelligence*, 6(1), 4–5.

OpenAI. (2019). *OpenAI Five defeats Dota 2 world champions*. https://openai.com/index/openai-five-defeats-dota-2-world-champions/

OpenAI. (2024, February 9). *Sora: Creating video from text*. https://openai.com/index/sora/

O'Reilly-Shah, V. N., Gentry, K. R., Walters, A. M., Zivot, J., Anderson, C. T., & Tighe, P. J. (2020). Bias and ethical considerations in machine learning and the automation of perioperative risk assessment. *British Journal of Anaesthesia*, 125(6), 843–846.

Otrel-Cass Kathrin and Fasching, M. (2021). Postdigital truths: Educational reflections on fake news and digital identities. In M. Savin-Baden (Ed.), *Postdigital humans: Transitions, transformations and transcendence* (pp. 89–108). Springer International Publishing. https://doi.org/10.1007/978-3-030-65592-1_6

Özdan, S. (2021). The right to freedom of expression versus legal actions against fake news: A case study of Singapore. In A. MacKenzie, J. Rose, & I. Bhatt (Eds.), *The epistemology of deceit in a postdigital era: Dupery by design* (pp. 77–94). Springer International Publishing. https://doi.org/10.1007/978-3-030-72154-1_5

Park, E., & Gelles-Watnick, R. (2023). Most Americans haven't used ChatGPT; few think it will have a major impact on their job. *Pew Research Center*, 28.

Patton, D. U., Frey, W. R., McGregor, K. A., Lee, F. T., McKeown, K., & Moss, E. (2020, February). Contextual analysis of social media: The promise and challenge of eliciting context in social media posts with natural language processing. In D. Slack., S. Hilgard., E. Jia., S. Singh, & H. Lakkaraju, H. *Proceedings of the AAAI/ACM Conference on AI, Ethics, and Society* (pp. 337–342). Association for Computing Machinery.

Paz, F. J., & Cazella, S. C. (2019). Academic analytics: a systematic review of literature. *International Journal of Development Research*, 9(11), 31710–31716.

Pearce, D. (1995). *Hedonistic imperative* (pp. 161–182). David Pearce.

Pelz, D. C., & Andrews, F. M. (1966). *Scientists in organizations: Productive climates for research and development.* John Wiley.

Perrotta, C., & Selwyn, N. (2019). Deep learning goes to school: Toward a relational understanding of AI in education. *Learning, Media and Technology*, 45(3), 251–269. https://doi.org/10.1080/17439884.2020.1686017

Peters, M. A. (2012). Bio-informational capitalism. *Thesis Eleven*, 110(1), 98–111.

Pranam, A. (2019, November 29). Why the retirement of Lee Se-Dol, former "Go" champion, is a sign of things to come. Forbes. https://www.forbes.com/sites/aswinpranam/2019/11/29/why-the-retirement-of-lee-se-dol-former-go-champion-is-a-sign-of-things-to-come/

Putnam, R. D. (2000). *Bowling alone: The collapse and revival of American community.* Simon Schuster.

Qiu, J. L. (2016). Goodbye iSlave: A digital manifesto. University of Illinois Press.

Quantic Dream. (2018). *Detroit: Become Human* [Video game]. Sony Interactive Entertainment.

Radu, R. (2021). Steering the governance of artificial intelligence: National strategies in perspective. *Policy and Society*, 40(2), 178–193. https://doi.org/10.1080/14494035.2021.1929728

Raffaghelli, J. E., Ferrarelli, M., & Rodríguez, N. L. (2025). Slowness as postdigital positionality in the era of generative AI: A conversation. *Postdigital Science and Education*, 1–26. https://doi.org/10.1007/s42438-025-00554-z

Ranjan, J., & Foropon, C. (2021). Big data analytics in building the competitive intelligence of organizations. *International Journal of Information Management*, 56, 102231.

Reader, J., & Evans, A. (2019). *Ethics after new materialism: A modest undertaking.* William Temple Foundation.

Reader, J., & Savin-Baden, M. (2020). Ethical conundrums and virtual humans. *Postdigital Science and Education*, 2(2), 289–301. https://doi.org/10.1007/s42438-019-00095-2.

Recchia, G. (2020). The fall and rise of AI: Investigating AI narratives with computational methods. In S. Dillon, S. Cave, & K. Dihal (Eds.), *AI narratives: A history of imaginative thinking about intelligent machines* (pp. 382–408). Oxford University Press.

Reicho, M., & Otrel-Cass, K. (2024). In pictures we trust: Evaluating digital information and disinformation with phenomenon-based learning in secondary schools: Visual knowledge creation and critique. *Video Journal of Education and Pedagogy, 1*(aop), 1–25.

Reiss, M. V. (2023). Testing the reliability of chatgpt for text annotation and classification: A cautionary remark. *arXiv preprint arXiv:2304.11085.*

Ring, L., Shi, L., Totzke, K., & Bickmore, T. (2015). Social support agents for older adults: Longitudinal affective computing in the home. *Journal on Multimodal User Interfaces, 9*, 79–88.

Robertson, A. (2024, February 21). Google apologizes for "missing the mark" after Gemini generated racially diverse Nazis. *The Verge.* https://www.theverge.com/2024/2/21/24079371/google-ai-gemini-generative-inaccurate-historical

Roden, D. (2014). *Posthuman life: Philosophy at the edge of the human.* Routledge.

Roddenberry, G. (1966, September 8). *Star Trek* [TV series]. NBC.

Rothblatt, M. A. (2014). *Virtually human: The promise—and the peril—of digital immortality.* Macmillan.

Rudolph, J., Tan, S., & Tan, S. (2023). ChatGPT: Bullshit spewer or the end of traditional assessments in higher education? *Journal of Applied Learning and Teaching, 6*(1), 1–22. https://doi.org/10.37074/jalt.2023.6.1.9

Rytting, C. M., Sorensen, T., Argyle, L., Busby, E., Fulda, N., Gubler, J., & Wingate, D. (2023). Towards coding social science datasets with language models. *arXiv preprint arXiv:2306.02177.*

Salmeri, A. (2020, December 5). Op-ed | No, Mars is not a free planet, no matter what SpaceX says. *SpaceNews.* https://spacenews.com/op-ed-no-mars-is-not-a-free-planet-no-matter-what-spacex-says/

Sandvig, C., Hamilton, K., Karahalios, K., & Langbort, C. (2016). When the algorithm itself is a racist: Diagnosing ethical harm in the basic components of software. *International Journal of Communication, 10.* https://ijoc.org.

Savin-Baden, M. (2010). *A practical guide to using Second Life in higher education.* McGraw-Hill Education (UK).

Savin-Baden, M., Bhakta, R., & Burden, D. (2016, May). Cyber enigmas? Passive detection and pedagogical agents: Can students spot the fake? In *Proceedings of the 10th International Conference on Networked Learning 2016* (pp. 456–463). Lancaster University.

Savin-Baden, M., Burden, D., & Taylor, H. (2017). The ethics and impact of digital immortality. *Knowledge Cultures, 5*(2), 178–196.

Savin-Baden, M., & Mason-Robbie, V. (Eds.). (2020). *Digital afterlife: Death matters in a digital age.* CRC Press.

Savin-Baden, M. (2022). *Digital afterlife and the spiritual realm.* Chapman and Hall/CRC

Savin-Baden, M. (2024). *Digital and postdigital learning for changing universities.* Routledge.

Savin-Baden, Z. (2025). Fighting a just war in a digital realm. In M. Power & M. Savin-Baden (Eds.), *Just war theory and artificial intelligence: Challengesa and consequences* (pp. 122–133). Routledge.

Saygin, A. P., Cicekli, I., & Akman, V. (2000). Turing test: 50 years later. *Minds and Machines*, 10(4), 463–518.

Scharre, P. (2023). *Four battlegrounds: Power in the age of artificial intelligence*. W. W. Norton & Company.

Segerstad, Y. H. a., Bell, J., & Yeshua-Katz, D. (2022). A sort of permanence: Digital remains and posthuman encounters with death. *Conjunctions*, 9(1), 1–12. https://doi.org/10.2478/tjcp-2022-0001

Shakespeare, W. (1600/2016). *King Henry IV Part 2: Third Series*. Bloomsbury Publishing.

Shanahan, M. (2024). Talking about large language models. *Communications of the ACM*, 67(2), 68–79.

Shankar, S., Halpern, Y., Breck, E., Atwood, J., Wilson, J., & Sculley, D. (2017). No classification without representation: Assessing geodiversity issues in open data sets for the developing world. *arXiv preprint arXiv:1711.08536*.

Sharples, M. (2022). Automated essay writing: An AIED opinion. *International Journal of Artificial Intelligence in Education*, 32(4), 1119–1126.

Shepherd, T. (2024, July 6). Real criminals, fake victims: How chatbots are being deployed in the global fight against phone scammers. *The Guardian*. https://www.theguardian.com/technology/article/2024/jul/07/ai-chatbots-phone-scams

Sherman, N. (2024, September 20). Three Mile Island nuclear site to reopen in Microsoft deal. *BBC News*. https://www.bbc.co.uk/news/articles/cx25v2d7zexo

Shilov, A. (2024, November 25). Microsoft says Word and Excel AI data scraping was not switched to enabled by default (Updated). *Tom's Hardware*. https://www.tomshardware.com/tech-industry/artificial-intelligence/microsoft-word-and-excel-ai-data-scraping-slyly-switched-to-opt-in-by-default-the-opt-out-toggle-is-not-that-easy-to-find

Shusterman, N. (2016). *Scythe*. Simon & Schuster.

Sigmund, T. (2021). Senseless Transhumanism. In W. Hofkirchner & H.- J. Kreowski (Eds.), *Transhumanism: The proper guide to a posthuman condition or a dangerous idea?* (pp. 65–78). Springer. https://doi.org/10.1007/978-3-030-56546-6_4.

Simon, M. (2009). HP looking into claim webcams can't see black people. *CNN*, December, 23.

Sinclair, C. (2021). Learning from the dupers: Showing the workings. In A. MacKenzie, J. Rose, & I. Bhatt (Eds.), *The epistemology of deceit in a postdigital era: Dupery by design* (pp. 233–249). Springer International Publishing.

Sircar, A. (2025, May 22). House bill barring state AI regulations sparks outcry. *Forbes*. https://www.forbes.com/sites/anishasircar/2025/05/22/total-control-to-ai-firms-us-house-bill-barring-state-oversight-draws-ire/

Smuha, N. A. (2024). *Algorithmic rule by law: How algorithmic regulation in the public sector erodes the rule of law*. Cambridge University Press.

Smuha, N. A., & Yeung, K. (2024). The European Union's AI Act: beyond motherhood and apple pie? Available at SSRN. Available at https://dx.doi.org/10.2139/ssrn.4874852

Smuha, N. A. (2025). An introduction to the law, ethics and policy of artificial intelligence. *The Cambridge handbook of the law, ethics and policy of artificial intelligence* (pp. 1–33). Cambridge University Press.

Snowden, E. (2014, July 18). Edward Snowden interview: The edited transcript. *The Guardian*. https://www.theguardian.com/world/2014/jul/18/-sp-edward-snowden-nsa-whistleblower-interview-transcript

Sparrow, R., & Zhang, E. (2025). (Re)Animating the ancestors: Digital personality emulations, ancestor veneration, and ethics. *New Media and Society*. Published online: 23 February 2025. https://doi.org/10.1177/14614448251317461

Steinhart, E. (2014). *Your digital afterlives: Computational theories of life after death*. Springer.

Starlink. (2024). *Starlink legal documents*. https://www.starlink.com/legal/documents/DOC-1020-91087-64

Statista. (2023a). Leading user concerns held about ChatGPT among respondents in Southeast Asia as of February 2023. https://www.statista.com/statistics/1382944/sea-top-user-concerns-about-chat-gpt/

Statista. (2023b). Perceived reliability of ChatGPT in South Korea as of March 2023. https://www.statista.com/statistics/1381444/south-korea-chatgpt-result-reliability-perception/

Statista. (2023c). Share of people who want to use ChatGPT in Japan as of June 2023. https://www.statista.com/statistics/1376113/japan-share-of-people-who-want-to-use-chatgpt/

Strubell, E., Ganesh, A., & McCallum, A. (2020, April). Energy and policy considerations for modern deep learning research. *Proceedings of the AAAI Conference on Artificial Intelligence, 34*(09), 13693–13696. https://doi.org/10.1609/aaai.v34i09.7123

Suarez-Villa, L. (2001). The rise of technocapitalism. *Science & Technology Studies, 14*(2), 4–20.

Suresh, H., & Guttag, J. (2021, October). A framework for understanding sources of harm throughout the machine learning life cycle. In *Proceedings of the 1st ACM Conference on Equity and Access in Algorithms, Mechanisms, and Optimization* (pp. 1–9). Association for Computing Machinery.

Ta, V. Griffith, C., Boatfield, C., Wang, X.,Civitello, M., Bader, H., DeCero, E., & Loggarakis, A., (2020). User experiences of social support from companion chatbots in everyday contexts: Thematic analysis. *Journal of Medical Internet Research, 22*(3), e16235.

Tangermann, V. (2023, December 5). You'll be astonished how much power it takes to generate a single AI image. *Futurism*. https://futurism.com/the-byte/power-generate-single-ai-image

Taras, D., & Steel, P. (2007). We provoked business students to unionize: Using deception to prove an IR point. *British Journal of Industrial Relations, 45*(1), 179–198.

Taylor, L., Meyer, E. T., & Schroeder, R. (2014). Bigger and better, or more of the same? Emerging practices and perspectives on big data analysis in economics. *Big Data and Society, 1*, 2053951714536877.

Tecno Mobile. (2025). *Universal Tone*. Retrieved from https://www.tecno-mobile.com/universaltone/

Terrell, M. (2023, November 28). A first-of-its-kind geothermal project is now operational. *Google Blog*. https://blog.google/outreach-initiatives/sustainability/google-fervo-geothermal-energy-partnership/

Thaldar, D. (2025). How effectively can ChatGPT-4 draft data transfer agreements for health research? *Humanities and Social Sciences Communications, 12*(1). https://doi.org/10.1057/s41599-025-04643-z

The Associated Press. (2025, April 18). Trump promotes lab leak theory on government COVID website. *AP News*. https://apnews.com/article/trump-covid-origin-lab-leak-fauci-c8767c1e2c5698c845059ab7f0534ff7

The Spiffing Brit. (2024, August 1). *This video breaks YouTube 4x* [Video]. YouTube. https://www.youtube.com/watch?v=29_0_wNR7yY

Thermo. (2024, July 1). *Nigel's unbelievable Minecraft mastery* [Video]. YouTube. https://www.youtube.com/watch?v=Sb4Q-cO4uBo

Thompson, T. (2017, October 31). *The perfect organism: The AI of Alien: Isolation.* Game Developer. https://www.gamedeveloper.com/design/the-perfect-organism-the-ai-of-alien-isolation

Tiku, N. (2022, June 11). The Google engineer who thinks the company's AI has come to life. *The Washington Post.*

Titus, L. M. (2024). Does ChatGPT have semantic understanding? A problem with the statistics-of-occurrence strategy. *Cognitive Systems Research, 83*, 101174.

Tolkien, J. R. R. (1955/1995). *Lord of the rings.* Harper Collins.

Turing, A. M. (1950). Computing machinery and intelligence. *Mind, 59*(236), 433–460.

Turvey, K. (2024). Review of Sarah Hayes, Michael Jopling, Stuart Connor, and Matthew Johnson (Eds.). (2023). *Human data interaction, disadvantage and skills in the community: Enabling cross-sector environments for postdigital inclusion. Postdigital Science and Education.* https://doi.org/10.1007/s42438-024-00452-w

Ulbricht, L., & Yeung, K. (2022). Algorithmic regulation: A maturing concept for investigating regulation of and through algorithms. *Regulation & Governance, 16*(1), 3–22. https://www.unoosa.org/oosa/en/ourwork/spacelaw/treaties/introouterspacetreaty.html

UN Environment Programme. (2024, September 21). AI has an environmental problem. Here's what the world can do about that. *UNEP.* https://www.unep.org/news-and-stories/story/ai-has-environmental-problem-heres-what-world-can-do-about

United Nations Office for Outer Space Affairs (UNOOSA). (1966). *The outer space treaty.* United Nations Office for Outer Space Affairs.

Vahl, S. (2024, November 28). Cloned customer voice beats bank security checks. *BBC News.* https://www.bbc.co.uk/news/articles/c1lg3ded6j9o

Vallis, C. (2025). Barbie meets generative AI in education: Neither artificial nor intelligent? *Educational Philosophy and Theory, 57*(10), 871–882.

Villaronga, E. F. (2019). "I Love You," said the robot: Boundaries of the use of emotions in human-robot interactions. In *Emotional design in human-robot interaction* (pp. 93–110). Springer.

Wakefield, J. (2021, February 26). MyHeritage offers 'creepy' deepfake tool to reanimate dead. *BBC News.* https://www.bbc.co.uk/news/technology-56210053

Waymo. (2024). *Safety Impact.* Waymo. https://waymo.com/safety/impact/

Wiederhold B. K. (2019). Can artificial intelligence predict the end of life... and do we really want to know? *Cyberpsychology, Behaviour and Social Networks, 22* (5), 297–299.

Yang, M.-H., Kriegman, D. J., & Ahuja, N. (2002). Detecting faces in images: A survey. *IEEE Transactions on Pattern Analysis and Machine Intelligence, 24*(1), 34–58.

Yeung, K. (2017) Algorithmic Regulation: A Critical Interrogation. *Regulation & Governance, 12*(4), 505–523.

Yoo, C.-m. (2019, November 27). *(Yonhap Interview) Go master Lee says he quits unable to win over AI Go players.* Yonhap News Agency. https://en.yna.co.kr/view/AEN20191127004800315

Yu, S., Carroll, F., & Bentley, B. L. (2024). Trust and trustworthiness: Privacy protection in the ChatGPT era. In *Data Protection: The Wake of AI and Machine Learning* (pp. 103–127). Springer Nature Switzerland.

Yuan, X., He, P., Zhu, Q., & Li, X. (2018). Adversarial examples: Attacks and defenses for deep learning. *arXiv*. https://arxiv.org/abs/1712.07107

Yuan, Y., Su, M., & Li, X. (2024, June). What makes people say thanks to AI. In *International Conference on Human-Computer Interaction* (pp. 131–149). Springer Nature Switzerland.

Zaini, R. M., Elmes, M. B., Pavlov, O. v., & Saeed, K. (2019). Organizational dissent dynamics in universities: Simulations with a system dynamics model. *Management Communication Quarterly*, *33*(3), 419–450. https://doi.org/10.1177/089331891 9846625

Zamen, W. (2009). *HP computers are racist*. YouTube. https://www.youtube.com/watch?v=t4DT3tQqgRM

Zhang, N., Duan, H., Guan, Y., Mao, R., Song, G., Yang, J., & Shan, Y. (2024). The 'Eastern Data and Western Computing' initiative in China contributes to its net-zero target. *Engineering*. https://doi.org/10.1016/j.eng.2024.08.010

Zhang, Z., Ning, H., Shi, F., Farha, F., Xu, Y., Xu, J., Zhang, F., & Choo, K. K. R. (2022). Artificial intelligence in cyber security: Research advances, challenges, and opportunities. *Artificial Intelligence Review*, *55*(2), 1029–1053.

Zhong, J., Zhong, Y., Han, M., Yang, T., & Zhang, Q. (2024). The impact of AI on carbon emissions: Evidence from 66 countries. *Applied Economics*, *56*(25), 2975–2989. https://doi.org/10.1080/00036846.2023.2288663

Zuboff, S. (2019). *The age of surveillance capitalism: The fight for a human future at the new frontier of power:* Profile Books.

Index

Note: **Bold** page numbers refer to tables and *italic* page numbers refer to figures.

For Product Safety Concerns and Information please contact our EU
representative GPSR@taylorandfrancis.com
Taylor & Francis Verlag GmbH, Kaufingerstraße 24, 80331 München, Germany